Lecture Notes in Mathematics

Edited by A. Dold and B. Eckmann

654

Joe P. Buhler

Icosahedral Galois Representations

Springer-Verlag
Berlin Heidelberg New York 1978

Author
Joe P. Buhler
Mathematics Department
The Pennsylvania State University
University Park, PA 16802/USA

Library of Congress Cataloging in Publication Data

Buhler, Joe P 1950-
Icosahedral galois representations.

(Lecture notes in mathematics ; 654)
Bibliography: p.
Includes index.
1. Algebraic number theory. 2. Galois theory.
3. Automorphic forms. I. Title. II. Series: Lecture notes in mathematics (Berlin) ; 654.
QA3.L28 no. 654 [QA247] 510'.8s [512'.74] 78-9714
ISBN 0-387-08844-X

AMS Subject Classifications (1970): 12A55, 12B10, 10D10

ISBN 3-540-08844-X Springer-Verlag Berlin Heidelberg New York
ISBN 0-387-08844-X Springer-Verlag New York Heidelberg Berlin

Printed in Germany

Printing and binding: Beltz Offsetdruck, Hemsbach/Bergstr.
2141/3140-543210

TABLE OF CONTENTS

TABLES

INTRODUCTION

It is conjectured that there is a natural one-to-one correspondence between certain modular forms of weight one for congruence subgroups of $SL_2(\mathbb{Z})$ and odd two dimensional galois representations $\rho:\mathrm{Gal}(\bar{\mathbb{Q}}/\mathbb{Q}) \longrightarrow GL_2(\mathbb{C})$. The L-series associated (by Hecke) to a modular form should agree with the L-series associated (by Artin) to the corresponding galois representation.

The following theorems show that this is indeed the case if the L-series of all odd two-dimensional galois representations are entire. For a description of the notation and a more detailed discussion of the situation the reader should consult [Serre, D] or [Deligne-Serre].

<u>Theorem L-W: Let</u> ρ <u>be an odd two-dimensional galois representation over</u> $\mathbb{Q}$ <u>with conductor</u> N <u>and determinant</u> ϵ. <u>If</u> $L(s,\rho\otimes\lambda)$ <u>is an entire function for all "twists"</u> $\rho\otimes\lambda$ <u>of</u> ρ <u>by one dimensional representations</u> λ, <u>then there is a modular form</u> f <u>of type</u> $(1,\epsilon,N)$ <u>such that</u> $L_f(s) = L(s,\rho)$.

<u>Theorem D-S: If</u> f <u>is a newform of type</u> $(1,\epsilon,N)$ <u>then there is an odd two dimensional galois representation</u> $\rho:\mathrm{Gal}(\bar{\mathbb{Q}}/\mathbb{Q}) \to GL_2(\mathbb{C})$ <u>with conductor</u> N <u>and determinant</u> ϵ, <u>such that</u> $L_f(s) = L(s,\rho)$.

Remarks: 1) Implicit in the statements of these theorems is an identification of Dirichlet characters $\epsilon:(\mathbb{Z}/N\mathbb{Z})^* \longrightarrow \mathbb{C}^*$ and characters (= one dimensional representations) $\epsilon:G_{\mathbb{Q}} \longrightarrow \mathbb{C}^*$. This equivalence comes from class field theory (over $\mathbb{Q}$), and one can view the conjectural correspondence between modular forms and galois representations as a natural generalization of this identification.

2) Theorem L-W is a consequence of some work of Langlands and Weil together with Langlands' results on the constants in the functional equation for $L(s,\rho)$. In fact the work of Jacquet and Langlands ([Jacquet-Langlands]) shows that this theorem is true in greater generality: one replaces $\mathbb{Q}$ with an arbitrary global field and considers the relationship between representations of the associated

Weil group and "admissible" representations of GL(2) of the adele group of the global field. If GL(2) is replaced with an arbitrary reductive group we arrive at the context for Langlands' vast generalization of the conjectural correspondence above.

3) Theorem D-S is proved in [Deligne-Serre]. If ρ is a galois representation coming from a newform via Theorem D-S then it follows from some results of Hecke that Artin's conjecture is true for ρ: $L(s, \rho)$ is an entire function. Thus there is a bijection between the newforms described in Theorem D-S and odd two-dimensional galois representations that, together with all of their twists, satisfy Artin's conjecture. Unfortunately there is no known way to construct all modular forms of weight one on $\Gamma_0(N)$ and there is no general way to determine whether a given galois representation satisfies Artin's conjecture. Nonetheless this is an interesting special case of the Langlands conjectures not only because of the existence of Theorem D-S, but, as we shall see, because it is possible to compute specific examples.

A two dimensional galois representation over $\mathbb{Q}$ determines a faithful two-dimensional representation of the galois group $Gal(K/\mathbb{Q})$ of a finite galois extension of $\mathbb{Q}$. The finite subgroups of $GL_2(\mathbb{C})$ were classified by Klein; their image in $PGL_2(\mathbb{C})$ must be either cyclic, dihedral, or the group of symmetries of one of the Platonic solids: the tetrahedral group (isomorphic to A_4), the octahedral group (isomorphic to S_4) or the icosahedral group (isomorphic to A_5). Any finite subgroup of $GL_2(\mathbb{C})$ is a cyclic central extension of one these projective groups.

The validity of Artin's conjecture for cyclic ($\Leftrightarrow$ reducible) or dihedral ($\Leftrightarrow$ monomial) two dimensional representations follows from the fact, due to Hecke, that abelian L-series are entire. Recently it was proved in [Langlands] that Artin's conjecture is true for tetrahedral representations and some octahedral representations. In all these cases the conjectural correspondence given in the first paragraph above is true. This leaves the case of icosahedral representations, which is the only case in which the associated galois groups $G(K/\mathbb{Q})$ are not solvable.

The original goal of this work was to establish Artin's conjecture for some suitably chosen icosahedral representations. The technique follows [Tate, N133]: one uses a specific galois representation to find a modular form of weight one and then applies Theorem D-S to get a galois representation that must satisfy Artin's conjecture.

In order to construct global icosahedral representations it was necessary to analyze carefully certain two-dimensional local galois representations. Everything done here goes over without serious difficulty to the case of representations of prime degree. These facts complement Koch's results on the structure of the images of primitive (= not induced) local representations of prime degree. The basic result here is a formula for the minimal conductor of a "lifting" of such a representation. (Koch has since extended his results to primitive representations of arbitrary degree).

The specific contents are as follows:

Chapter 1 is devoted to the problem of "lifting" projective galois representations to linear representations of Weil groups of a class formation. This should be contrasted with the approach in [Serre, D] which considers galois representations.

Chapter 2 is concerned with the problem alluded to above: finding the minimal conductor of a lifting of a primitive local galois representation of prime degree.

Chapter 3 applies the ideas of the first two chapters to two-dimensional galois representations over $\mathbf{Q}$. A search for icosahedral representations of low conductor is then described; the results of this search are contained in Table A_5 at the end of these notes. The lowest conductor found was $800 = 2^5\ 5^2$.

Chapter 4 describes the calculation of the L-series of an icosahedral representation. These ideas are applied to the icosahedral representation of conductor 800. This calculation involves some computations in a sextic extension of

$\mathbf{Q}$ that are related to the computation of a generalized ideal class group. These computations require a computer and lead to some interesting algorithms; however the results can be checked by hand computation.

Chapter 5 recalls some results on "classical" modular forms of weight one and then uses these results to do some linear algebra that is necessary for the last chapter.

The last chapter contains the main result: There is an icosahedral form of level 800. This verifies a specific instance of the initial conjectural correspondence described above and in the process establishes Artin's conjecture for the associated representation. This is probably the first known instance of the validity of Artin's conjecture for a non-solvable galois representation that is not a positive linear combination of monomial representations. Following a suggestion of Serre it is then easy to check that all of the L-series attached to non-trivial representations of the underlying A_5 extension are entire (appendix 7).

Some of the details of the proof are inevitably computational; if the reader wishes to verify the proof completely, he/she should have access to a computer. Fortunately the calculations are of the sort that can be subjected to stringent consistency checks; for instance, all of the systems of linear equations considered are vastly over-determined so that the existence of any solution at all is a minor miracle. All of the computations were done on a PDP-11 equipped with a UNIX operating system ([Thompson-Ritchie]). The value of this operating system (and accompanying language) can not be overestimated; the computations here would require so much more time and patience on many systems as to be impractical.

In general the auxiliary results are developed only to the depth necessary for the proof of the existence of an icosahedral form. For instance, the results in chapter 2 on local galois representations can be generalized to primitive representations of arbitrary degree and to a large class of imprimitive representations. The results in appendix 6 on fourier expansions at all cusps will appear (in a slightly more general context and with proofs) in a joint paper with Jim Weisinger.

There are a number of appendices, some of which function almost as coroutines (as opposed to subroutines) and thereby form an indispensable part of some of the proofs. The tables for a given chapter are collected at the at the end of that chapter.

I would like to thank the following people for their help and encouragement: Brent Byer, Dick Gross, Gunter Harder, Peter Langston, Winnie Li, Barry Mazur, Victor Miller, Judy Moore, Jean-Pierre Serre, Michael Shia, and Tucker Taft. Some assistance of special importance came from Jim Weisinger, who worked on the problem of making some complicated formulae simplify, and from Michael Penk, who worked on the problem of making a computer do hard things quickly.

This volume is essentially my thesis, with only minor changes. My advisor was John Tate, whose assistance and enthusiasm contributed more to the results herein than can be adequately acknowledged.

CHAPTER ONE

Let F be a local or a global field, $\bar{F}$ a separable algebraic closure of F, and $G_F = G(\bar{F}/F)$ the galois group of $\bar{F}$ over F. A linear galois representation over F is a continuous homomorphism

$$\mathbb{T}: G_F \longrightarrow GL(V).$$

where V is a finite dimensional complex vector space. Such a representation $\mathbb{T}$ determines a projective galois representation $T:G_F \longrightarrow PGL(V)$ by composition with the map $GL(V) \longrightarrow PGL(V)$. If $\mathbb{C}^*$ is identified with the group of homotheties of V then we get a commutative diagram

1)
$$\begin{array}{ccccccccc} & & & & G_F & & & & \\ & & & \mathbb{T} \swarrow & & \searrow T & & & \\ 0 & \longrightarrow & \mathbb{C}^* & \longrightarrow & GL(V) & \longrightarrow & PGL(V) & \longrightarrow & 0 \end{array}$$

in which the horizontal row is exact. If $\lambda:G_F \longrightarrow \mathbb{C}^*$ is a character (= a one-dimensional representation) and $\lambda \otimes \mathbb{T}$ denotes the representation defined by $(\lambda \otimes \mathbb{T})(g) = \lambda(g)\mathbb{T}(g)$ for $g \in G_F$, then $\lambda \otimes \mathbb{T}$ determines the same projective representation T. We call $\lambda \otimes \mathbb{T}$ the twist of $\mathbb{T}$ by λ.

If T is obtained from $\mathbb{T}$ by the above procedure then $\mathbb{T}$ is said to be a lifting of T. If $T:G_F \longrightarrow PGL(V)$ is a fixed projective representation then the obstruction to lifting T is an element of the cohomology group $H^2(G_F,\mathbb{C}^*)$. This group is trivial and hence liftings of a given projective galois representation always exist (for more details and a general discussion of this situation see [Serre, D]).

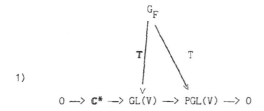

In this chapter we consider the problem of lifting a projective galois representation to a representation of the Weil group associated to F. In order to analyze this problem we briefly recall some notation concerning Weil groups. To fix the ideas we consider only the Weil groups and class formations attached to local or global fields; but it would be easy to prove the same results for an abstract topological class formation satisfying suitable axioms.

If K is any finite galois extension of F then put

$C_K = K^*$ if K is local

$C_K =$ the idele class group of K if K is global

Class field theory gives an injection

$$\mathrm{inv}_F : H^2(G_F, \varinjlim C_K) \longrightarrow \mathbf{Q}/\mathbf{Z} ,$$

where the inductive limit is over all finite extensions K in $\overline{F}$. For each such K there is a "fundamental class"

$$\xi(K/F) \in H^2(G(K/F), C_K)$$

whose invariant is $\mathrm{inv}_F(\xi(K/F)) = 1/[K:F]$. If F is not an archimedean local field then inv_F is an isomorphism.

The fundamental class $\xi(K/F)$ determines a group extension

$$C_K \longrightarrow W(K/F) \longrightarrow G(K/F) .$$

These "Weil" groups W(K/F) obey various functorial properties; for instance, there is a canonical isomorphism of C_F with $W(K/F)^{ab}$. For a discussion of the properties of the Weil groups W(K/F) see [Artin-Tate, ch. 15], or [Weil, CF].

By taking the inverse limit over all finite extensions K in $\overline{F}$ we get the Weil group W_F. This group is equipped with a natural topology in which it is locally compact; moreover the canonical map $W_F \longrightarrow G_F$ has a dense image. The topology on W_F is such that any continuous representation $W_F \longrightarrow GL(V)$ factors through a representation of $W(K/F)$ for some finite extension K of F.

Let $T:G_F \longrightarrow PGL(V)$ be a projective galois representations that factors through a representation of a finite galois group $G(K/F)$. Since $G(K/F)$ is a quotient of W_F we can think of T as a projective representation of W_F. A <u>lifting</u> of T is then a linear representation $\mathbb{T}:W_F \longrightarrow GL(V)$ whose associated projective representation is T.

If $\mathbb{T}$ is a lifting of T then the restriction of $\mathbb{T}$ to $W(\overline{F}/K) \cong W_K$ takes its values in the group of homotheties ($= \ker(GL(V) \longrightarrow PGL(V))$). Since this group is isomorphic to $\mathbb{C}^*$ this restriction can be identified with a quasicharacter

$$\chi':W_K \longrightarrow \mathbb{C}^*.$$

Since W_K^{ab} is isomorphic to C_K, χ' uniquely determines a quasicharacter

$$\chi:C_K \longrightarrow \mathbb{C}^*$$

which we will call the <u>centric character</u> of $\mathbb{T}$. Since $PGL(V)$ acts trivially on $\mathbb{C}^*$ the quasicharacter χ is invariant under the action of $G(K/F)$; i.e. $\chi^s = \chi$ for all $s \in G(K/F)$, where by definition $\chi^s(x) = \chi(x^s)$.

If $\chi:C_K \longrightarrow \mathbb{C}^*$ is the centric character for <u>some</u> lifting of T then we will say that χ is <u>centric for</u> T.

To what extent do χ and T determine $\mathbb{T}$? If $\mathbb{T}$ and $\mathbb{T}'$ are two liftings with the same centric character χ then $\mathbb{T}' = \lambda \otimes \mathbb{T}$ where $\lambda \in Hom(W_F,\mathbb{C}^*)$ is trivial on W_K. Therefore λ factors through $W_F/W_K \cong G(K/F)$. Thus the centric character determines the lifting up to the relatively harmless operation of twisting by elements of the finite group $Hom(G(K/F),\mathbb{C}^*)$. If, for example, $G(K/F)$ were simple, then a lifting of T would be completely specified by giving a centric character for T.

Recall that if K/F is any finite galois extension with galois group $G = G(K/F)$ then the cohomology group $H^{-1}(G,C_K)$ is defined to be C_K^N/C_K^I where

$$C_K^N = \ker N_{K/F} = \{ x \in C_K : N_{K/F}(x) = 1 \}$$

$$C_K^I = (IG)C_K = \text{the group generated by all } x^s/x, \quad x \in C_K, \ s \in G.$$

Let $c(T) \in H^2(G(K/F),\mathbf{C}^*)$ be the pullback via T of the cohomology class in $H^2(PGL(V),\mathbf{C}^*)$ of the exact sequence in diagram 1. Let $\delta:H^n(G,\mathbf{C}^*) \longrightarrow H^{n+1}(G,\mathbf{Z})$ be the coboundary map associated with the sequence

$$0 \longrightarrow \mathbf{Z} \longrightarrow \mathbf{C} \longrightarrow \mathbf{C}^* \longrightarrow 0$$

(where the map from $\mathbf{C}$ to $\mathbf{C}^*$ is the mapping $x \longrightarrow \exp(2\pi i x)$). Then the projective representation T determines a homomorphism from C_K^N to $\mathbf{Q}/\mathbf{Z}$ defined by taking a class $[x] \in H^{-1}(G,C_K)$, cupping with $\delta c(T)$ and then applying the invariant map inv_F:

$$H^{-1}(G,C_K) \times H^3(G,\mathbf{Z}) \longrightarrow H^2(G,C_K) \longrightarrow \mathbf{Q}/\mathbf{Z}.$$

This is a key ingredient in the following theorem.

Theorem 1: *Let* $T: W_F \longrightarrow PGL(V)$ *be a projective representation over a local or global field* F *that factors through a representation of the galois group* $G = G(K/F)$ *of a finite extension* K *of* F. *Let* $c = c(T) \in H^2(G,\mathbf{C}^*)$ *be the class associated to* T *as explained above. A quasicharacter*

$$\chi: C_K \longrightarrow \mathbf{C}^*$$

is centric for T *if and only if it satisfies the following condition:*

A) $\chi(x) = \exp(2\pi i \, inv_F([x] \cup \delta c((T))))$ *for all* $x \in C_K^N$.

Remarks:

1) The hypothesis that T factors through a representation of a finite galois group is true for an important class of projective representaions of W_F; namely those projective representations T which have a lifting $\mathbb{T}$ which is irreducible and not induced from a representation of a proper subgroup of W_F. Indeed, if such a $\mathbb{T}$ factors through W(K/F) for a finite galois extension K of F, then a standard argument (see [Serre, RG, p.77]) shows that $\mathbb{T}$ takes the abelian group C_K to the $\mathbb{C}^*$, so that T factors through G(K/F).

2) Condition A) is reminiscent of the standard characterization of the reciprocity map in terms of the invariant map and characters of the galois group.

3) Note that if χ satisfies A) then it is trivial on K^{*I} so that

$$\chi(x^{s-1}) = \chi^{s-1}(x) = \chi^{s}(x)/\chi(x) = 1 , \ s \in G, \ x \in K^*,$$

and χ is automatically invariant under the action of G.

Proof: Suppose that a quasicharacter χ is given. The problem of finding a lifting $\mathbb{T}$ whose centric character is χ is exactly the problem of completing the following diagram:

$$\begin{array}{ccccccccc}
0 & \longrightarrow & \mathbb{C}^* & \longrightarrow & GL(V) & \longrightarrow & PGL(V) & \longrightarrow & 0 \\
 & & \uparrow \chi & & \vdots\, \mathbb{T} & & \uparrow T & & \\
0 & \longrightarrow & C_K & \longrightarrow & W(K/F) & \longrightarrow & G & \longrightarrow & 0
\end{array}$$

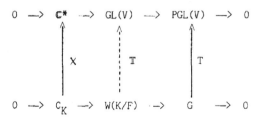

It is a standard result from the cohomology of groups (e.g. [Artin-Tate, p.179]) that this diagram can be completed if and only if χ is a G-homomorphism, and takes the cohomology class of the second row into the pullback of the cohomology class of the first row to G. Since PGL(V) acts trivially on $\mathbf{C}^*$ the condition that χ be a G-homomorphism just says that $\chi^s = \chi$ for all $s \in G$; so from now on we assume that the χ under consideration has this property.

Let $\xi = \xi(K/F)$ be the fundamental class of K/F so that the cohomology class of the lower horizontal row above is just ξ. By definition the pullback of the cohomology class of the upper row to G is $c = c(T)$. Thus T admits a lifting with centric character χ if and only if

B) $\quad \chi_*(\xi) = c.$

Now we will show that A) and B) are equivalent.

Let $x \in C_K$ be of norm 1 so that x determines a cohomology class

$$[x] \in H^{-1}(G,C_K) = C_K{}^N/C_K{}^I \quad .$$

The cup product theorem of class field theory says that cupping by the fundamental class ξ gives an isomorphism

$$H^{-3}(G,\mathbf{Z}) \cong H^{-1}(G,C_K) \quad .$$

Hence there is an $\alpha \in H^{-3}(G,\mathbf{Z})$ such that

$$\alpha \cup \xi = [x]$$

Assume that condition B) holds. Apply χ_* to the preceeding equation and use B) to get:

$$[\chi(x)] = \chi_*([x]) = \chi_*(\alpha \cup \xi)$$

$$= \alpha \cup \chi_*(\xi) = \alpha \cup c \quad \in \quad H^{-1}(G,\mathbf{C}^*)$$

Apply the coboundary operator to both sides to get

$$\delta[\chi(x)] = \delta(\alpha \cup c) = \alpha \cup \delta c \in H^0(G,\mathbf{Z})$$

Cup with the fundamental class and recall the definition of α:

$$\xi \cup \delta[\chi(x)] = \xi \cup \alpha \cup \delta c = [x] \cup \delta c \in H^2(G,C_K).$$

Therefore

$$\delta[\chi(x)]/n = \mathrm{inv}_F([x] \cup \delta c) \in \quad (1/n)\mathbf{Z}/\mathbf{Z} \subset \mathbf{Q}/\mathbf{Z}$$

where $n = [K:F]$. By tracing through the definitions one sees that the inverse of the isomorphism

$$\mu_n(\mathbf{C}) \cong H^{-1}(G,\mathbf{C}^*) \xrightarrow{\delta} H^0(G,\mathbf{Z}) \cong \mathbf{Z}/n\mathbf{Z}$$

is the map $x \longrightarrow \exp(2\pi i x/n)$ so that we have

$$\chi(x) = \exp(2\pi i\ \mathrm{inv}_F([x] \cup \delta c)) \in \mu_n(\mathbf{C}) \subset \mathbf{C}^*.$$

This is A) of the theorem.

Reversing the argument, we find that A) implies that

$$\alpha \cup \chi_*(\xi) = \alpha \cup c \quad \text{for all } \alpha \in H^{-3}(G,\mathbf{Z}).$$

Since the pairing

$$H^{-3}(G,\mathbf{Z}) \times H^2(G,\mathbf{C}^*) \longrightarrow H^{-1}(G,\mathbf{C}^*) \cong \mu_n(\mathbf{C})$$

is a perfect duality, we conclude that A) implies B). This finishes the proof of the theorem.

Let χ_T be the character of C_K^N given by the right hand side of A) in Theorem 1. The theorem then says that a quasicharacter χ of C_K is centric for T if and only if it extends χ_T.

<u>Corollary: Any projective representation of G_F has a lifting to a representation of W_F.</u>

Proof: C_K^N is a closed subgroup of C_K, and any character of a closed subgroup of a locally compact abelian group has an extension to the whole group. q.e.d.

If $T: G_F \longrightarrow PGL(V)$ is as in the above theorem, then a lifting $\mathbf{T}: W_F \longrightarrow GL(V)$ is said to be a <u>galois lifting</u> if it factors through a representation of the quotient G_F of W_F. It is easy to check that an arbitrary lifting is a galois lifting if and only if its centric character $\chi: C_K \longrightarrow \mathbb{C}^*$ comes from a character of G_K^{ab}.

As indicated above, any projective representation $T: G_F \longrightarrow PGL(V)$ has a galois lifting ([Serre, D]). For local fields this is an immediate consequence of the above theorem. Indeed, if K is a local field then any element of K^* of norm 1 is a unit, and any quasicharacter of K^* can be multiplied by an appropriate unramified quasicharacter to give a character of finite order.

If F is a global function field then a similar argument again shows that any projective galois representation has a galois lifting.

Suppose that F is a number field. Then a quasicharacter χ of C_K corresponds to a character of G_K if it is trivial on the connected component D_K. If it is possible to extend χ_T to C_K in such a way that the resulting quasicharacter vanishes on D_K then

$$D_K \cap C_K{}^N \subset \mathrm{kernel}(\chi_T) \quad .$$

This condition is in fact sufficient (see the lemma below). Since the odd cohomology groups, and in particular the H^{-1}, of the connected component are trivial we have

$$D_K{}^N = D_K{}^I .$$

It follows that the above condition is satisfied, so that it is possible to choose the centric character χ to be trivial on D_K. This argument is really the same as the argument using the fact that $H^2(G_F, \mathbb{C}^*) = \{0\}$, since the key point in the proof in [Serre, D] that $H^2(G_F, \mathbb{C}^*) = \{0\}$ is the triviality of the odd cohomology of the connected component.

The sufficiency of the above condition is a special case of the following standard fact:

Extension Lemma: *Let G be an abelian locally compact toplogical group, B be a closed subgroup, K a compact subgroup, and χ' a character of K. Then χ' can be extended to a continuous homomorphism*

$$\chi : G \longrightarrow \mathbb{C}^*$$

with $\chi(B) = \{1\}$ if and only if

$$B \cap K \subset \mathrm{kernel}\ \chi' \quad .$$

In the context of the argument above the connected component D_K played the role of the closed subgroup, and the kernel of the norm was the compact subgroup. The pattern of this argument - namely using the extension lemma in tandem with condition A) of Theorem 1 - will be used frequently. Two more illustrations of this follow; one is a theorem that will be useful in both local and global contexts, and the other is an application to the problem of finding the minimal order of a character that is centric for a projective representation T.

Theorem 2: *Let $T:G_F \longrightarrow PGL(V)$ be a projective galois representation over a local or global field F, with image G(K/F). Let c(T) be the cohomology class in $H^2(G(K/F),\mathbb{C}^*)$ determined by T, and let E be a field that is intermediate between F and K. Then there exists a quasicharacter $\psi:C_E \longrightarrow \mathbb{C}^*$ such that $\psi \circ N_{K/E}$ is centric for T if and only if*

$$\mathrm{Res}^{G(K/F)}_{G(K/E)}(c(T)) = 0.$$

Remark: The group cohomology restriction map

$$\mathrm{Res}: H^2(G(K/F),\mathbb{C}^*) \longrightarrow H^2(G(K/E),\mathbb{C}^*)$$

corresponds to restricting T to G_E; more precisely:

$$\mathrm{Res}(c(T)) = c(T|G_E).$$

Proof: Since $N_{K/E}:C_K \longrightarrow C_E$ is an open map a character χ is of the form $\psi \circ N_{K/E}$ if and only if χ vanishes on the kernel of $N_{K/E}$. The extension lemma says that χ_T can be extended to C_K with this property if and only if

$$\mathrm{kernel}(N_{K/E}) \subset \mathrm{kernel}(\chi_T).$$

If $N_{K/E}x = 1$ then x determines cohomology classes in both $H^{-1}(G(K/E),C_E)$ and $H^{-1}(G(K/F),C_F)$; with the obvious notation we have $\mathrm{Cores}([x]_{K/E}) = [x]_{K/F}$.

Assume that Res (c(T)) = 0. Then

$$\mathrm{inv}_F([x]_{K/F} \cup \delta c(T)) = \mathrm{inv}_E([x] \cup \mathrm{res}(\delta c(T))) = 0$$

so that by condition A) of Theorem 1, $\chi_T(x) = \chi(x) = 1$. This shows that any centric character χ must vanish on the kernel of $N_{K/E}$ and hence must be of the desired form.

The injectivity of inv_E together with the fact that the cup product establishes a perfect duality between H^{-1} and H^3 shows that if χ is of the form $\psi \circ N_{K/E}$ then $\mathrm{res}(\delta c(T))$ must be trivial. Thus χ is of the above form if and only if $\mathrm{res}(c(T))$ is 0 as claimed in the theorem. q.e d

If $T:G_F \longrightarrow PGL(V)$ is a projective galois representation and $\mathbf{T}:G_F \longrightarrow GL(V)$ is a galois lifting, then the image of $\mathbf{T}$ is a central extension of the image of T by a finite cyclic group. Call the order of this group the index of the lifting, so that the index of a galois lifting is just the order of the centric character χ.

A character of C_K is of order dividing m if and only if it is trivial on C_K^m. Thus we are in a position to apply the extension lemma to the closed subgroup C_K^m and the character χ_T. The result is that T has a lifting of index dividing m if and only if $\chi_T(C_K{}^m \cap C_K{}^N) = 1$, which is that same as

*) $\qquad x \in C_K,\ (N_{K/F}x)^m = 1 \quad \Rightarrow \quad \chi_T(x^m) = 1.$

Now we take K and F to be nonarchimedean local fields. Let T be of degree n and let n' be the order of the largest n-power root of unity in F. Then c(T) is of order dividing n in $H^2(G(K/F, \mathbb{C}^*)$ and hence χ_T is of order dividing n.

<u>Proposition: If F is a nonarchimedean field then any projective galois representation has a lifting of index at most</u> nn'.

Proof: Apply the condition *) above. If $N_{K/F}(x^{nn'}) = 1$ then $N_{K/F}(x^{n'}) = 1$ since an n-power root of unity in F is of order at most n'. Thus $\chi_T(x^{n'})^n = 1$ since χ_T is of order n. q.e.d.

If F is a nonarchimedean local field of residue characteristic 2 then the condition *) is in [Weil, ED, p. 15]. For the S_4 and A_4 representations over $\mathbf{Q}_2$ the minimal index is either 2 or 4. This index will be determined in chapter 3 by global means; the resulting calculations are considerably easier in general than the computations required by the application of the above condition. Note that in the function field case (characteristic F= 2) the above analysis shows that there is always a lifting of index 2. Warning: the lifting of minimal index is usually not the lifting of minimal conductor. In the next chapter the above ideas will be used to solve the problem of minimizing the conductor of a lifting of a given projective representation of prime degree.

CHAPTER TWO

An irreducible continuous finite dimensional representation $\mathbb{T}:G \longrightarrow GL(V)$ of a topological group G is said to be primitive if it is not induced from a representation of a proper open subgroup of finite index in G. Thus $\mathbb{T}$ is imprimitive (= properly induced) if and only if there is decomposition

$$V \cong V_1 \oplus \ . \ . \oplus \ V_n, \qquad \text{with } n > 1, \text{ and } V_i \neq \{0\} \text{ for all } i$$

such that for all $g \in G$

$$V_i \longrightarrow \mathbb{T}(g)V_i$$

is a permutation of the V_i. Since homotheties take any subspace V_i to itself the primitivity or imprimitivity of a representation $\mathbb{T}$ depends only on the underlying projective representation $T:G \longrightarrow PGL(V)$. Similarly the reducibility/irreducibility of a representation depends only on the underlying projective representation.

If $T:G \longrightarrow PGL(V)$ is a projective representation of a finite group G then a <u>lifting</u> of T is a representation $\mathbb{T}:G' \longrightarrow GL(V)$ such that:

1) G' is a central extension of G

2) $\mathbb{T}$ takes the kernel of $G' \longrightarrow G$ to the group of homotheties of V

3) the representation $G \longrightarrow PGL(V)$ induced by $\mathbb{T}$ is the original projective representation T.

It is well known that such a lifting $\mathbb{T}$ always exists for any T ([Dornhoff, p. 135]). By the argument given above, the primitivity or imprimitivity of a lifting $\mathbb{T}$ is determined by the representation T. Thus we say that a projective representation is primitive if some (and hence any) lifting is primitive. Similarly we can speak of reducible or induced projective representations (even though it is not possible to take direct sums of projective representations or to induce a projective representation).

Lemma 1: Any non-trivial irreducible projective representation of a supersolvable group is imprimitive.

Proof: A central extension of a supersolvable group is supersolvable, and an irreducible linear representation of a supersolvable group is monomial ([Serre, RG, p. 82]. q.e.d.

The object of this chapter is to study local galois representations. In principle, reducible and imprimitive representations can be analyzed in terms of representations of smaller degree, so it is natural to begin by considering primitive representations.

Let F be a non-archimedean local field of residue characteristic p. Then the galois group G_F contains the wild ramification subgroup P_F which is a pro-p-group. The quotient G_F/P_F is the galois group of the maximal tame extension of F and is therefore a (profinite) supersolvable group.

Theorem 3: Let $\mathbb{T}:G_F \longrightarrow GL(V)$ be a primitive galois representation. Then the restriction of $\mathbb{T}$ to P_F is irreducible.

Proof: According to a theorem of Clifford (see [Serre, RG, p. 77] or [Dornhoff, p. 72]) the primitivity of $\mathbb{T}$ implies that the restriction of $\mathbb{T}$ to P_F is isotypic. Let $\mathbb{T}_1$ be the irreducible representation of P_F that occurs in the restriction of $\mathbb{T}$ to P_F, and let T and T_1 be the projective representations corresponding to $\mathbb{T}$ and $\mathbb{T}_1$. In this situation another theorem of Clifford (see [Dornhoff, p. 185]) says that

$$T = \overline{T}_1 \otimes T_2$$

where $\overline{T}_1$ is an extension of T_1 to all of G_F and T_2 is trivial on P_F.

Suppose that the conclusion of the theorem is false, so that $\mathbb{T}|P_F$ is reducible and the degree of T_1 is strictly smaller than the degree of T. Then T_2 is a representation of degree bigger than one. But T_2 is then an imprimitive representation since it is an irreducible representation of a supersolvable group. This contradicts the assumed primitivity of $\mathbb{T}$ because

$$\mathbb{T}_1' \otimes \mathrm{Ind}(\mathbb{T}_2') \cong \mathrm{Ind}(\mathrm{Res}(\mathbb{T}_1') \otimes \mathbb{T}_2')$$

so that the tensor product of an induced representation with another representation is again induced; similarly for projective representations. This shows that $\mathbb{T}|P_F$ must be irreducible.

<u>Corollary: The degree of</u> $\mathbb{T}$ <u>is a power of</u> p.

Proof: Irreducible representations of p-groups are of p-power order. q.e.d.

Remark: The proof of the theorem actually shows that if $\mathbb{T}$ is a primitive representation of a group G and H is a normal subgroup of G such that G/H is supersolvable, then the restriction of $\mathbb{T}$ to H is irreducible.

Thus the simplest non-trivial primitive representations of G_F are those of prime degree p. If $T: G_F \longrightarrow PGL_p(\mathbb{C})$ is a primitive representation then we will let $G = T(G_F)$, resp. $P \doteq T(P_F)$, be the image of G_F, resp. P_F. In this context we will let K and F' be the extensions of F that correspond to G and P so that

$$G = G(K/F) \qquad P = G(K/F').$$

The p-group P is the wild ramification group of G, and F' is the maximal tame extension of F inside K. Put $H = G/P = G(F'/F)$.

It is well known (see [Dornhoff, p.181 and p.185]) that irreducible finite projective p-groups that are normal subgroups of primitive projective groups are elementary abelian groups whose order is the square of the degree. Thus P is an elementary abelian group of order p^2 on which H = G/P acts, so that T determines a two-dimensional representation of H over $\mathbf{F}_p$. By using results of Suprenenko on solvable linear groups one can give a precise description of the possible groups H and the corresponding projective representations T (see [Suprenenko] or [Koch, DP]).

In order to give this classification it is convenient to introduce some notation. Let e be the ramification index of F'/F and let f be the residue class degree. Identify P with a fixed two-dimensional $\mathbf{F}_p$ vector space so that a projective representation T determines a mapping from H to $GL_2(\mathbf{F}_p)$. If A is (the $\mathbf{F}_p$-points of) a nonsplit torus in $SL_2(\mathbf{F}_p)$ then let N(A) be the normalizer of A in $SL_2(\mathbf{F}_p)$; A is a cyclic group of order p+1 contained as a subgroup of index two in N(A), and the quotient $N(A)/A \cong \{\pm 1\}$ acts on A by inversion. If $p \neq 2$ then N(A) is a "generalized quaternion" group.

Proposition 1: *Let* $T: G_F \longrightarrow PGL_p(\mathbf{C})$ *be a primitive representation such that* F' *is (as above) the maximal tame extension of* F *fixed by the kernel of* T. *Then* T *determines (the isomorphism class of) a representation*

$$H \longrightarrow SL_2(\mathbf{F}_p)$$

of H = G(F'/F) *that is faithful and irreducible and whose image lies in the normalizer* N(A) *of a nonsplit torus* A. *Any subgroup of* N(A) *whose intersection with* A *has order bigger than* 2 *can arise in this way.*

In order to explicitly count the possible representations T it is necessary to know the possible finite groups H, the number of tame extensions F'/F with galois group H, the number of possible representations of H as above, the number of fields K for a given F'/F, and the number of representations for a given K and given representation of H.

Addendum to Proposition 1 ([Koch, DP]):

In all cases the tame ramification index e divides p+1.

1A) Suppose that p = 2. Then H is either of order 3 or order 6 (cyclic or isomorphic to the symmetric group S_3). The representation of H into $SL_2(\mathbf{F}_2) \cong S_3$ is unique and the group G is isomorphic to A_4 or S_4.

1B) Suppose that p is odd. If the residue field of F is an extension of $\mathbf{F}_p$ of even degree then H is cyclic of order dividing p+1. If the residue field is of odd degree over $\mathbf{F}_p$ then either

a) H is cyclic of order dividing p+1 and e = 1 or e = 2

b) f = 2 and e is even and greater than 2.

c) f = 4 and e is odd and greater than 2.

2) The group H admits $\varphi(ef)/2$ representations as in 1) of Proposition 1 above (here φ is the euler phi function).

3) If H is noncyclic then there is a unique tame extension F' of F with galois group H. If H is cyclic then for each pair of integers e and f whose product is the order of H then there are $e_1 \varphi(e_2)$ such tame extensions, where e_1 is the largest divisor of e that is relatively prime to f.

4) If F is of characteristic 0, then for each field F' there are

$(p^{2m}-1)/(p-1)$ if H is cyclic

$(p^{2m}-1)/(p^2-1)$ if H is noncyclic

fields K such that G(K/F) admits a faithful primitive representation of degree p, where $m = [F:\mathbf{Q}_p]$.

5) For each representation of H = G(F'/F) as above and each field extension K/F'

there are

$(p-1)/2$	*if H is noncyclic and p is odd*
$p-1$	*otherwise*

primitive representations T that determine K and the given representation of H.

Remark: Koch has considered the structure of primitive representations of arbitrary degree in [Koch, CP] and gives analogous results in this case.

Corollary: The Schur multiplier group of P (resp. G) is cyclic of order p and the restriction map

$$H^2(G,\mathbb{C}^*) \longrightarrow H^2(P,\mathbb{C}^*)$$

is an isomorphism.

Proof: The usual technique for calculating the Schur multiplier of abelian groups ([Huppert, p.652]) shows that $H^2(P,\mathbb{C}^*)$ is of order p. Let $c \in H^2(PGL_p(\mathbb{C}),\mathbb{C}^*)$ be the canonical class determined by $GL_p(\mathbb{C})$. Then in fact one can use an explicit description of the map $P \longrightarrow PGL_p(\mathbb{C})$ to show that the pullback of c to $H^2(P,\mathbb{C}^*)$ is nontrivial, and therefore generates this group.

Suppose p is odd. Then by the Proposition P is the p-sylow subgroup of G and so the restriction map is an injection on the p-primary part of $H^2(G,\mathbb{C}^*)$. The q-sylow subgroups of G, for $q \neq p$, are either cyclic or generalized quaternion (the latter is possible only if q= 2). These groups have trivial Schur multiplier ([Huppert, p.643]), so $H^2(G,\mathbb{C}^*)$ is a p-group. Since $H^2(G,\mathbb{C}^*)$ is a non-trivial group (if there is a lifting $G \to GL_p(\mathbb{C})$ that lifts $G \to PGL_p(\mathbb{C})$ then we could restrict to P to get a lifting $P \to GL_p(\mathbb{C})$ which would contradict the statement above about the cohomology class determined by $P \to PGL_p(\mathbb{C})$), all of the assertions of the corollary are true for odd p.

For $p = 2$, G is either A_4 or S_4 and the corollary can be proved by direct calculation. q.e.d.

If M/L is any galois extension of nonarchimedean local fields let $b_0(M/L)$ be the last "break" in the filtration of $G(M/L)$ by the lower numbering of the ramification subgroups; i.e. $b_0(M/L)$ is the largest integer n such that

$$G(M/L)_n \neq \{0\}.$$

Since a non-trivial cyclic subgroup of P is not invariant under conjugation by G/P (since G/P acts irreducibly) we see that the cyclic subgroups of P of order p are not normal in G. Hence if $r = b_0(K/F)$ then

$$P = G_1 = G_2 = \ . \ . \ . = G_r \neq G_{r+1} = \{0\}.$$

The rest of the notation used here for local fields is more standard: $e_{M/L}$ is the ramification index, $f_{M/L}$ is the residue class degree, $\mathfrak{P}_L$ is the prime ideal in the ring of integers of L, $\mathfrak{D}_{M/L}$ is the different, v_L is the valuation on L such that $v_L(L^*) = \mathbf{Z}$, U_L^n is the group of units of the ring of integers of L such that $v_L(x-1) \geq n$, and $\varphi_{M/L}$ is the Herbrand function.

If $\mathbf{T}: G_F \longrightarrow GL_n(\mathbf{C})$ is a galois representation over F then let $a(\mathbf{T}) \in \mathbf{Z}$ be the (exponent of the) Artin conductor of $\mathbf{T}$. If $\mathbf{T}$ is a one-dimensional representation which corresponds by class field theory to a character $\lambda: F^* \longrightarrow \mathbf{C}^*$, then $a(\mathbf{T})$ is the usual conductor of λ; namely the smallest integer n such that λ is trivial on U_F^n. The conductor $a(\mathbf{T})$ can be defined by requiring that $a(\mathbf{T}_1 \oplus \mathbf{T}_2) = a(\mathbf{T}_1) + a(\mathbf{T}_2)$ and putting

$$a(\mathbf{T}) = (1 + \sup\{m: \mathbf{T}(G_F{}^m) \neq 1\})/n.$$

for irreducible $\mathbf{T}$, where $G_F{}^m$ is the m-th ramification group in the <u>upper</u> numbering.

If T is a projective representation of G_F then define A(T) to be the smallest possible value of a($\mathbb{T}$) as $\mathbb{T}$ ranges over all liftings of T.

The object of the remainder of this chapter is to prove the following theorem.

Theorem 4: Let $T:G_F \longrightarrow PGL_p(\mathbb{C})$ be a primitive projective galois representation of degree p over a local field of residue characteristic p. Then

$$A(T)= p + \left(\frac{p+1}{e}\right)b_0(K/F),$$

where K is the fixed field of the kernel of T, and e is the tame ramification degree of K/F.

We start by reducing the theorem to the problem of minimizing the possible conductors of characters $\chi:K^* \longrightarrow \mathbb{C}^*$ that are centric for T.

Proposition 2: Let L/F be a finite galois extension with galois group $\mathbb{G} = G(L/F)$, and let $\mathbb{T}:\mathbb{G} \longrightarrow GL_n(\mathbb{C})$ be a faithful irreducible representation. Let $r = b_0(L/F)$ and assume that there is a subgroup H of $\mathbb{G}_r$ such that $\mathbb{T}(H)$ is a nontrivial subgroup of the group of scalar matrices. Let E be a galois extension of F in L that is fixed by H. Then

$$e_{E/F}a(\mathbb{T})= n\, v_E(\mathcal{D}_{E/F}) + a(\mathbb{T}|G(L/E)).$$

Remarks:

1) The key hypothesis of the theorem, namely the existence of H, is very weak; often $H = \mathbb{G}_r$ works. This is the case if $\mathbb{T}$ is primitive, because then $\mathbb{G}_r$, which is a normal abelian subgroup of $\mathbb{G}$, goes into the group of scalar matrices under $\mathbb{T}$. More generally, if the restriction of $\mathbb{T}$ to the wild ramification group $\mathbb{G}_1$ is irreducible then $\mathbb{G}_r$, which is a central subgroup of $\mathbb{G}_1$ ([Serre, CL, p. 77]), goes

into the group of scalar matrices by Schur's Lemma.

2) Putting $\mathrm{cond}(\mathbb{T}) = \mathfrak{p}^{a(\mathbb{T})}$, we can write conclusion of the proposition as

$$\mathrm{cond}(\mathbb{T}) = (\mathfrak{D}_{E/F})^{n}\mathrm{cond}(\mathbb{T}|G(L/E)).$$

Proof: Let $\Theta_{\mathbb{T}}$ be the character of $\mathbb{T}$ and let $\Theta_{\mathbb{G}}$ be the character of the Artin representation of $\mathbb{G}$ ([Serre, CL, p. 107]). Then

$$a(\mathbb{T})= (1/\#G) \sum_{g \in \mathbb{G}} \Theta_{\mathbb{T}}(g)\ \Theta_{\mathbb{G}}(g)$$

where $\#G$ is the number of elements of $\mathbb{G}$.

Let χ be the centric character of $\mathbb{T}$, which here we think of as a faithful character defined on the cyclic subgroup of $\mathbb{G}$ that goes, via $\mathbb{T}$, into the scalar matrices. Since

$$\Theta_{\mathbb{T}}(xy)= \mathrm{trace}(\mathbb{T}(xy))= \chi(x)\Theta_{\mathbb{T}}(y), \text{ for } x \in H,\ y \in \mathbb{G}$$

and since $\Theta_{\mathbb{G}}$ is constant on non-trivial cosets $C = gH$, $g \notin H$, we have

$$\sum_{g \in C} \Theta_{\mathbb{T}}(g)\ \Theta_{\mathbb{G}}(g) = \text{constant} \sum_{x \in H} \chi(x) = 0$$

and therefore

$$\sum_{g \in \mathbb{G}} \Theta_{\mathbb{T}}(g)\ \Theta_{\mathbb{G}}(g) = \sum_{g \in H'} \Theta_{\mathbb{T}}(g)\ \Theta_{\mathbb{G}}(g) .$$

for any subgroup H' that contains H.

If we take $H'= G(L/E)$ in the above formula and use the fact that

$$\Theta_{\mathbb{G}}|H = f_{E/F}\ (v_E(\mathfrak{D}_{E/F})r_H + \Theta_H)$$

(see [Serre, CL, p. 108]; r_H is the regular representation of H), then a straightforward computation gives the statement in the proposition. q.e.d.

Corollary: Let T, K, and e be as in Theorem 4. Let $\mathbb{T}$ be any lifting of T, and let $\chi : K^* \longrightarrow \mathbb{C}^*$ be the centric character of $\mathbb{T}$. Put $r = b_0(K/F)$. Then

$$pe\ a(\mathbb{T}) = rp^2 + ep^2 - r - 1 + a(\chi).$$

Proof: Apply the proposition with $H = \mathbb{G}_r$ (possible by the remark after the proposition) and E = K; note that $e_{K/F} = p^2e$ and $v_K(\mathfrak{D}_{K/F}) = p^2e-1+r(p^2-1)$. q.e.d.

Corollary: Let $\mathbb{T}$ be a lifting of T, and let $\mathbb{T}'$ be the restriction of $\mathbb{T}$ to P_F. Then

$$e\ a(\mathbb{T}) = e(p-1) + a(\mathbb{T}').$$

Proof: This formula follows immediately from the proposition by letting E be the maximal tame extension of F in K. q.e.d.

Remark: The last corollary is in [Weil, ED, p.19].

A little algebra now shows that Theorem 4 is an immediate consequence of the following assertion about centric characters.

Theorem 4': With the above notation, the smallest conductor of a character χ that is centric for T is

$$a(\chi) = (p+1)b_0(K/F)+1.$$

Using the ideas developed in chapter 1 we will now describe the minimal possible value of $a(\chi)$ in terms of the filtration on $H^{-1}(G,K^*)$ induced by the usual filtration on K^*. Since

$$H^{-1}(G,K^*) \cong H^{-3}(G,\mathbf{Z}) \cong H^3(G,\mathbf{Z}) \cong H^2(G,\mathbf{C}^*) \cong H^2(P,\mathbf{C}^*)$$

the index $[K^{*N}:K^{*I}]$ is equal to p. The cohomology class c(T) is nontrivial so the character χ must restrict to a nontrivial character on K^{*N} whose kernel is exactly K^{*I}. According to Theorem 1 and the Extension Lemma of the first chapter, it is possible to find a χ of conductor n if and only if

1) $$U_K^n \cap K^{*N} \subset K^{*I}.$$

This inclusion is true for sufficiently large n since for large enough n the group U_K^n is contained in a cohomologically trivial subgroup of K^*. Alternatively the existence of such an n follows from the existence of a lifting and hence some character χ centric for T.

For the remainder of this chapter we will let $b_{-1}(K/F)$ be the largest integer n that does <u>not</u> satisfy 1), with similar notation for other galois extensions of local fields.

Thus the minimal value of $a(\chi)$ for χ a character centric for T is $b_{-1}(K/F)+1$. Recall that F' is the maximal tame extension of F in K (so that F' is the fixed field of P). Theorem 4' follows from the following two assertions:

<u>Claim 1:</u> $b_{-1}(K/F) = b_{-1}(K/F')$.

<u>Claim 2:</u> $b_{-1}(K/F') = (p+1)b_0(K/F')$.

By using the discussion above and the corollaries to Proposition 2 it can be seen that claims are equivalent to the following assertions about liftings of T and

$T|P_F$:

<u>Claim 1': Let T' be the restriction of T to P_F. Then</u>

$$e\,A(T) = p(e-1) + A(T').$$

<u>Claim 2':</u> $A(T') = p+(p+1)b_0(K/F')$.

First we prove Claim 1 (using Claim 2) and then prove a slight generalization of Claim 2'.

By the corollary to Proposition 1, the restriction map

$$\mathrm{Res}:H^2(G,\mathbb{C}^*) \longrightarrow H^2(P,\mathbb{C}^*)$$

is an isomorphism. It follows by class field theory and duality that the corestriction map

$$\mathrm{Cor}:H^{-1}(P,K^*) \longrightarrow H^{-1}(G,K^*).$$

is an isomorphism. If we let NP be the norm map $N_{K/F'}$, and IP be the augmentation ideal for P so that $H^{-1}(P,K^*) \cong K^{*NP}/K^{*IP}$ then

$$K^{*NP} \cap K^{*I} = K^{*IP}.$$

by the injectivity of the corestriction map. Thus if x is in $K^{*NP} - K^{*IP}$ then x is in $K^{*N} - K^{*I}$ so that $b_{-1}(K/F) \geq b_{-1}(K/F')$.

The reverse inequality is a consequence of the following lemma.

<u>Lemma: Let</u> M/L <u>be a tame extension of local fields of residue characteristic</u> p. <u>Then</u>

$$H^{-1}(G(M/L),U_M^n) = 0$$

<u>for</u> $n > 0$.

Proof: Let H be a p-Sylow subgroup of G(M/L). Since U^n is a pro p-group the restriction map

$$\mathrm{Res}:H^{-1}(G(M/L),U_M^n) \longrightarrow H^{-1}(H,U_M^n)$$

is an injection. The tame ramification index of G(M/L) is prime to p, so H corresponds to an unramified extension M/M'. Thus the proof of the proposition is reduced to the case in which M/L is unramified.

The unramified case follows from the fact that U_M^n/U_M^{n+1} can be identified with the residue field, and the fact that the residue field is cohomologically trivial (this argument is essentially the proof of Lemma 2 in [Serre, CL, p. 193]). q.e.d

Now we can finish the proof of Claim 1 by showing that $b_{-1}(K/F') \geq b_{-1}(K/F)$. Suppose that $n > b_{-1}(K/F')$ and that $x \in (U_K^n)^N$. By the above lemma

2) $$N_{K/F'}x = \bar{y}^{s-1}\bar{z}^{t-1}$$

where $\bar{y}, \bar{z} \in U_{F'}^m$, $m = \langle \varphi_{K/F'}(n) \rangle$, and s, t are elements of G that generate $G/P \cong G(F'/F)$ (notation: $\langle m \rangle$= the smallest integer $\geq m$). By Claim 2, $n > r = b_0(K/F')$ so that $\bar{y}$ and $\bar{z}$ are norms from K:

$$\bar{y} = N_{K/F'}y,\ \bar{z} = N_{K/F'}z, \quad \text{with } y,z, \in U_K^n.$$

Let $x' = xy^{1-s}y^{1-t}$. Then 2) above implies that $N_{K/F'}x' = 1$. Since $x' \in U_K^n$, $n > b_{-1}(K/F')$ we must have $x \in K^{*I}$ which shows that $n > b_{-1}(K/F)$. This proves that $b_{-1}(K/F') \geq b_{-1}(K/F)$ which finishes the proof of Claim 1 above.

The crux of Theorem 4 above is contained in Claim 2' which is an assertion about the restriction T' of T to the wild ramification group P_F. Let E be an extension of F' of degree p in K. By theorem 2 a centric character for T' can be put in the form $\psi \circ N_{K/E}$ and it is easy to verify that $\mathrm{Ind}(\psi)$ is a lifting of T and that any lifting is of this form. For later purposes it is convenient to consider more general induced representations. The notation will be slightly altered (the field F' above will become the new ground field F and T is replaced by its restriction to the wild ramification group). But it is clear that Claim 2' follows as a special case of the following proposition by taking K/E to be of degree p, and assuming that there is only one "break" in the ramification filtration of G(K/F). It can be shown that this last assumption implies that

$$b_0(K/F) = b_0(K/E) = b_0(E/F)$$

so that (in the notation of the proposition) $r' = r+1$.

<u>Proposition 3:</u> *Let F be a local field of residue characteristic p and let E be a galois extension of F of degree p. Let $\psi_0 : E^* \to \mathbb{C}^*$ be a wildly ramified character such that $\mathbb{T} = \mathrm{Ind}(\psi_0)$ is an irreducible (p-dimensional) representation of G_F. Let T be the underlying projective representation and let K/F be the kernel field of T. Define r and r' by*

$$r = \begin{cases} 0 & \text{if } E/F \text{ is unramified} \\ b_0(E/F) & \text{if } E/F \text{ is ramified} \end{cases}$$

r' = the least integer n such that $U_E^n \subset N_{K/E}K^*$.

Then the minimal possible value of $a(\psi)$ for $\psi : E^* \to \mathbb{C}^*$ such that $\mathrm{Ind}(\psi)$ lifts T is

$$a(\psi) = r + r'$$

so that

$$A(T) = f_{E/F}(r+r') + v_E(\mathfrak{D}_{E/F}).$$

Proof: The case in which E/F is unramified follows the lines of the argument below; the proof is simpler and will be omitted.

If $\lambda : F^* \longrightarrow \mathbf{C}^*$ is a character of F^* then let $\lambda_E = \lambda \circ N_{E/F}$ be the character of E^* obtained by composing with the norm map.

If $\psi : E^* \longrightarrow \mathbf{C}^*$ is a character such that $\mathrm{Ind}(\psi)$ lifts T then $\mathrm{Ind}(\psi_0)$ and $\mathrm{Ind}(\psi)$ are twists of each other so that there is a character $\lambda : F^* \longrightarrow \mathbf{C}^*$ such that

$$\mathrm{Ind}(\psi) \cong \lambda \otimes \mathrm{Ind}(\psi_0) \cong \mathrm{Ind}(\lambda_E \psi_0).$$

The character of $\mathrm{Ind}(\psi)$ vanishes outside G_E since E/F is normal. The restriction of $\mathrm{Ind}(\psi)$ to G_E is of the form

$$\Sigma\, \psi^g$$

where the summation is over all elements g of the cyclic group G(E/F). Thus the projectivizations of $\mathrm{Ind}(\psi)$ and $\mathrm{Ind}(\psi_0)$ are equal if and only if

$$\psi = \lambda_E (\psi_0)^g$$

for some g in G(E/F). Since $a(\psi) = a(\psi^g)$ it follows that the proposition is reduced to the following assertion:

$$\inf a(\lambda_E \psi_0) = r + r'.$$

where the infimum is over all characters λ of F^*.

Let s be a generator of G(E/F), and let $\theta : E^* \longrightarrow \mathbb{C}^*$ be defined by

$$\theta(x) = \psi_0(x^s/x) = \psi_0^{\,s-1}(x).$$

Since K corresponds to the projectivization T of $\mathrm{Ind}(\psi_0)$ it follows that $N_{K/E}K^*$ is the intersection of the kernels of the conjugates of θ under G(E/F). Thus the conductor of θ is exactly r'. Note that θ kills F* so that r' is not congruent to 1 mod p.

Since G(E/F) is cyclic a character $\psi : E^* \longrightarrow \mathbb{C}^*$ is of the form $\lambda_E\psi_0$ if and only if

$$\psi^{s-1} = \theta = \psi_0^{\,s-1}$$

which is to say that ψ and ψ_0 agree on $\ker N_{E/F} = E^{*s-1}$. Thus we have a fixed character α on E^{*s-1} defined by

*) $$\alpha(x^{s-1}) = \theta(x)$$

(which is well defined since θ kills F*) and we want to find the extension of α to E* that has the smallest possible conductor. In order to apply the Extension Lemma of chapter 1 the following technical lemma is necessary.

Lemma: Let M/L be a wildly ramified galois extension of local fields of residue characteristic p. Assume that G(M/L) is cyclic of order p generated by s, and let $r = b_0(M/L)$. Any $y \in M^*$ can be written in the form

$$y = z\pi^m y_1 y_2 y_3 \cdot \,.\,.$$

where $z \in L^*$, $0 \leq m < p$, π is a uniformizing parameter in M, $y_i = 1$ if $i \equiv 0$ mod p, and for $i \not\equiv 0$ mod p either $y_i = 1$ or $y_i \in U_M^i - U_M^{i+1}$. If y has this form then

$$v_M(y^{s-1}-1) = r, \quad \underline{\text{if}}\ m \neq 0$$
$$= i+r, \quad \underline{\text{if}}\ m = 0\ \underline{\text{and}}\ i\ \underline{\text{is the smallest subscript for which}}\ y_i \neq 1.$$

Proof: It is easy to see that any y can be represented in the above form. If we apply s to this expression for y, and then divide by y we get

$$y^{s-1} = (\pi^{s-1})^m \prod y_i^{s-1} \quad .$$

By the assumptions above π^{s-1} must have the form 1 + a where $v_M(a) = r$. It is a standard fact that if y = 1 + b where $v_M(b) = j>0$ then

$$y^{s-1} \equiv 1 + jab \bmod \pi^{j+r+1}$$

and the last of assertion of the lemma follows. q.e.d.

Now we return to the context preceding the lemma; namely we want to find the minimal conductor of a character ψ that extends α. According to the previous lemma,

$$U_E^{r+r'} \cap E^{*s-1} \subset \ker(\alpha),$$

because of the definition of α and the fact that θ kills F* and $U_E^{r'}$. By the Extension Lemma it follows that we can choose a ψ extending α in such a way that $a(\psi) \leq r+r'$.

Choose a y in $U_E^{r'-1}$ with $\theta(y) \neq 1$, so that $\alpha(y^{s-1}) \neq 1$. By the previous lemma, $y^{s-1} \in U_E^{r+r'-1}$ (because of the fact that, as observed above, r'-1 is not divisible by p). It follows that $a(\psi) \geq r+r'$ for any ψ extending α. This shows that r+r' is the smallest possible value of $a(\psi)$ for a ψ such that Ind(ψ) is a twist of $\mathrm{Ind}(\psi_0)$, and finishes the proof of the proposition. q.e.d.

The device of using the Extension Lemma is efficient, but it has the drawback of being non-constructive. Thus the proof provides no explicit construction of the centric character χ of minimal conductor. It would also be nice to know the indices and determinants of the associated liftings. In the following chapters it will become necessary to grapple with some of these questions in the case of the unique A_4 extension of $\mathbb{Q}_2$.

CHAPTER THREE

Let $T:G_F \longrightarrow PGL_2(\mathbf{C})$ be a two-dimensional projective representation over a global field F. For each place v of F choose a decomposition group $G_v \longrightarrow G_F$. Then

$$G_v \cong G_{F_v} ,$$

so that the restriction of T to G_v is a local projective representation, T_v. Suppose that for each v we are given a lifting $\mathbb{T}_v$ of T_v. Is it possible to find a global lifting $\mathbb{T}$ whose restrictions to the G_v are the $\mathbb{T}_v$? In general this is far too much to hope for, but for $F = \mathbf{Q}$ there is an affirmative answer to a weaker question.

Theorem 5 (see [Serre, D]): Let $T:G_{\mathbf{Q}} \longrightarrow PGL_2(\mathbf{C})$ be a two dimensional projective representation with local components T_p for each rational prime p. For each p let $\mathbb{T}'_p$ be a lifting of T_p. Then there is a unique lifting $\mathbb{T}$ of T such that

$$\mathbb{T}_p = \mathbb{T}'_p \text{ on the inertia group } I_p \longrightarrow G_p.$$

Proof: Let $\mathbb{T}$ be any lifting of T. Then

$$\mathbb{T}_p = \lambda_p \otimes \mathbb{T}'_p$$

where λ_p is a character of G_p. Since the ground field is $\mathbf{Q}$ it is possible to find a unique Dirichlet character λ that agrees with the λ_p on the inertia subgroup of G_p, modulo the usual identification of Dirichlet characters with characters of $G_{\mathbf{Q}}$. The lifting $\lambda^{-1} \otimes \mathbb{T}$ now satisfies the conditions of the theorem and is unique because of the uniqueness of λ. q.e.d.

The conductor of a local representation depends only on the restriction of the representation to the inertia group. The above theorem therefore says that there is a well-defined minimal possible conductor of a lifting of a given projective representation over $\mathbf{Q}$. Let $T: G_{\mathbf{Q}} \to PGL_2(\mathbf{C})$ be a projective galois representation over $\mathbf{Q}$ and let cond(T) be the minimal conductor of a lifting of T. Let $A(T_p) = a(\mathbb{T}_p)$ be the (exponent of the) minimal conductor of a lifting of the local representation T_p. According to the above theorem the ideal cond(T) is

$$\mathrm{cond}(T) = \prod p^{A(T_p)} \quad .$$

The projective representation T also determines the parity of the determinant of any lifting. If $\mathbb{T}$ is a lifting of T then twisting $\mathbb{T}$ by a Dirichlet character changes $\det(\mathbb{T})$ by the square of that character; thus $\det(\mathbb{T})(-1) \in \{\pm 1\}$ depends only on T. If p is a place of $\mathbf{Q}$ and $\mathbb{T}_p$ is a lifting of T_p the define

$$(T,p) = \det(\mathbb{T}_p)(-1) \in \{\pm 1\}$$

(where $\mathbb{T}_p$ is any lifting of T_p). Note that (T,p) depends only on T_p. It is a straightforward exercise to verify that (T,p) can be obtained by pulling back the cohomology class of

$$\pm 1 \to SL_2(\mathbf{C}) \to PSL_2(\mathbf{C}) = PGL_2(\mathbf{C})$$

to $G_{\mathbf{Q}_p}$ via T_p, and then identifying $H^2(G_{\mathbf{Q}_p}, \pm 1)$ with $\{\pm 1\}$. Thus (T,p) should be thought of as an element of order 2 in the Brauer group $Br(\mathbf{Q}_p)$. It also follows from the definition that $\prod (T,p) = 1$ where the product is over all places of $\mathbf{Q}$ (including the infinite one). Thus if one knows (T,p) for all but one p then the remaining (T,p) is uniquely determined.

If λ is any Dirichlet character and λ_p is its "p-th component" then define

$$(\lambda,p) = \lambda_p(-1).$$

Again this can be interpreted cohomologically; (λ,p) is the element of order 2 in the Brauer group obtained by pulling back the cohomology class of

$$\pm 1 \longrightarrow \mathbb{C}^* \longrightarrow \mathbb{C}^*$$

(where the second arrow is the map taking x to x^2) to $G_{\mathbb{Q}_p}$ via λ_p. Also $\prod(\lambda,p) = 1$.

Theorem 6: *Let T be a two-dimensional galois representation over* $\mathbb{Q}$, *and let* λ *be a Dirichlet character. Then T has a lifting whose determinant is* λ *if and only if*

$$(T,p) = (\lambda,p) \text{ for all } p.$$

Remarks: By the fact that the product of all (T,p) is one it suffices to verify the condition of the theorem for all but one prime. Theorems 5 and 6 are due to Tate; for an exposition of these theorems and further consequences see [Serre, D].

Now let T be an icosahedral representation, so that the image of T is isomorphic to the alternating group A_5 on five symbols. In concrete terms this is an A_5 extension K of $\mathbb{Q}$ together with an embedding of $G(K/\mathbb{Q})$ into $PGL_2(\mathbb{C})$. The object of this chapter is to compute $A(T_p)$ and (T,p) for the various possible local representations T_p. The two conjugacy classes of embeddings of A_5 into $PGL_2(\mathbb{C})$ differ by an automorphism of $\mathbb{C}$ so cond(T) depends only on K; in fact $A(T_p)$ and (T,p) depend only on the extension K_p of $\mathbb{Q}_p$ determined by T_p. Since we are only interested here in the quantities $A(T_p)$ and (T,p) we will (for the remainder of this chapter) identify two galois representations that differ by an automorphism of $\mathbb{C}$.

The A_5 extension K of $\mathbf{Q}$ is conveniently specified by giving a quintic polynomial whose splitting field is K. In order to find $A(T_p)$ it is necessary to be able to recognize the local representation from the behavior of the quintic polynomial and to then have a table of the minimal conductors associated to the various local projective representations.

This information is tabulated in Table 3.1. There are 19 possible ramified local representations T_p (up to automorphisms of $\mathbf{C}$). Table 3.1 contains the following:

1) The sequence of ramification groups at p, starting with the decomposition group; C_n denotes the cyclic group of order n and D_n denotes the dihedral group of order 2n. Note that D_2 is the klein group of order 4 and D_3 is the symmetric group on 3 symbols.

2) the splitting of p in the root field of the quintic; $2^2 1$ means that p is the square of a prime of degree 2 times a prime of degree 1, etc.

3) the exponent of p in the discriminant of the root field of the quintic

4) $A(T_p)$

5) necessary conditions on p for this local representation to occur

Two dimensional projective groups are either cyclic, dihedral or primitive. In the following outline of how Table 3.1 was obtained the dihedral case is further subdivided into even or odd cases depending on whether or not the order of $G(K_p/\mathbf{Q}_p)$ is divisible by 4.

Cyclic

In this case the representation T_p is reducible. The projective representation corresponds to a character, say χ, of $\mathbb{Q}_p$. A reducible representation $\chi_1 \oplus \chi_2$ is a lifting of T_p if and only if $\chi_1/\chi_2 \cong \chi$. Since $a(\chi_1 \oplus \chi_2) = a(\chi_1) + a(\chi_2)$ it is clear that the minimal conductor is obtained if, for instance, one takes $\chi_1 = \chi \quad \chi_2 = 1$. Thus $A(T_p)$ is the conductor of χ, and $(T,p) = \chi(-1)$. Note that if T_p is unramified then $\mathbf{T}_p$ can be chosen to be unramified.

Odd Dihedral

In this case K_p is an extension of $\mathbb{Q}_p$ of degree 2n, n odd. The cyclic subgroup of $G(K_p/\mathbb{Q}_p)$ of order n fixes a quadratic extension, say E, of $\mathbb{Q}_p$. If $\psi: E^* \longrightarrow \mathbb{C}^*$ is a character that corresponds to the cyclic extension K_p/E and s is a generator of $G(E/\mathbb{Q}_p)$ then $\psi^s = \psi^{-1}$.

The Schur multiplier group of an odd dihedral group is trivial [Huppert, p. 646] so $G(K_p/\mathbb{Q}_p)$ admits a faithful two dimensional representation. The representation $\mathrm{Ind}(\psi, G_{\mathbb{Q}_p}/G_E)$ is such a representation. Twisting this representation by a character $\lambda: \mathbb{Q}_p \longrightarrow \mathbb{C}^*$ amounts to multiplying ψ by the character $\lambda \circ N_{E/\mathbb{Q}_p}$. If ψ agrees with $\lambda \circ N_{E/\mathbb{Q}_p}$ on x and x^s then

$$\psi(x) = \lambda(N_{E/\mathbb{Q}_p}(x)) = \lambda(N_{E/\mathbb{Q}_p}(x^s)) = \psi(x^s) = \psi(x)^{-1}$$

so $\psi(x) = 1$ (ψ is of odd order). Therefore the conductor of ψ can not be decreased by twisting by a character of the form $\lambda \circ N_{E/\mathbb{Q}_p}$ and the representation $\mathrm{Ind}(\psi)$ has the minimal conductor among the liftings of the original projective representation. The determinant of the induced representation $\mathrm{Ind}(\psi)$ is

$$\Theta\, \psi|_{\mathbb{Q}_p}$$

where Θ is the quadratic character associated to $E/\mathbb{Q}_p$, and so $(T,p) = \Theta(-1)$ (because ψ has odd order).

The only odd dihedral groups inside A_5 are of order 10 and 6; i.e. in our context n= 3 or n= 5.

Even Dihedral

The Schur multipler group of an even dihedral group is of order two ([Huppert, p.646]) so now we must use the theory of chapter 2. Fortunately the only even dihedral group contained in A_5 is the klein group of order 4. In this case the last proposition in chapter 2 is particularly easy to apply since the numbers r and r' are completely determined by the knowledge of the discriminant of the field and the residue class degree. The case of larger dihedral groups, which would occur if we were concerned with, for instance, S_4 extensions, appears to be genuinely more difficult; an example is calculated in appendix 4.

Again let $\psi: E^* \longrightarrow \mathbb{C}^*$ be such that Ind(ψ) lifts T_p, where E is a quadratic extension of $\mathbb{Q}_p$ that corresponds to a character $\Theta: \mathbb{Q}_p^* \longrightarrow \pm 1$. Let s be the non-trivial element of $G(E/\mathbb{Q}_p)$. It is easy to check that the cyclic extension K_p/E corresponds to the character

$$\psi^{s-1}: E^* \longrightarrow \mathbb{C}^*.$$

By the formula for the determinant of an induced representation

$$(T,p) = \Theta(-1)\,\psi(-1) = \Theta(-1)\,\psi^{s-1}(\sqrt{u})$$

where $E = \mathbb{Q}_p(\sqrt{u})$. Note that the last term on the right hand side depends only on K_p/E and not on the choice of ψ.

Primitive

In view of the corollary to Theorem 3 the primitive groups can occur only if $p = 2$. Since the groups A_4 and S_4 are solvable it is possible to use class field theory to enumerate the extensions of $\mathbb{Q}_2$ with these galois groups. There is a unique A_4 extension of $\mathbb{Q}_2$ and there are three S_4 extensions of $\mathbb{Q}_2$; some information about these fields is collected in Table 3.2. Of course only the A_4 field occurs as a subgroup of A_5; the corresponding local field will play a prominent role in the later chapters.

The information in Table 1 makes it possible to search for A_5 fields of "small" conductor. This was down by first sieving through a large number of quintic polynomials, eliminating first those whose discriminants were not squares, then the reducible polynomials and finally eliminating those whose galois groups were proper subgroups of A_5. For each surviving quintic the ring of integers of the root field was then computed. This determined the behavior of the ramified primes (and consequently the corresponding local representations) and the information in Table 3.1 could be applied to obtain the minimal conductor associated to any specific quintic polynomial. The smallest conductors that were obtained are listed in Table 3.3. Table A_5 contains all quintics that were found to have "conductor" less than 10000; the notation of that appendix is the same as the notation for Table 3.3. The remainder of this chapter consists of some remarks on the above procedure and the results obtained.

1. The sieving process above was applied to about 10^7-10^8 polynomials. Perhaps it is unfortunate that such a crude approach is necessary, but there is of course no known way to generate all A_5 extensions. The "systematic" procedures for finding A_5 extensions seem to invariably produce quintics of very large conductor.

2. Using some ideas of [Hunter] it is possible to do the search "effectively" in that it is possible to find a constant B such that any quintic field of discrim-

inant $< B$ was examined in the search. By using these ideas it is possible to obtain $B = 200^2$ as a very conservative estimate; i.e. any quintic whose root field has discriminant less than 40000 was included in the above search.

3. The irreducibility of a quintic $f(x)$ was tentatively checked by attempting to find a prime modulo which the quintic was irreducible; if no small prime was forthcoming then an attempt was made to factor the polynomial. The density of primes for which an A_5 quintic is irreducible is 5/12, and in fact any A_5 quintic found here was irreducible modulo a prime less than 20.

4. The galois group of an irreducible quintic polyomial with square discriminant is either C_5, D_5, or A_5. An initial attempt was made to verify that a given quintic had A_5 as its galois group by trying to find a prime, not dividing the discriminant of the quintic, modulo which the quintic factored as a cubic times two distinct linear polyomials. The density of such primes for an A_5 polyomial is 1/3, and since the "generic" irreducible quintic with square discriminant has galois group A_5 this test was usually successful. The obvious case where failure was expected was when the square root of the discriminant was the discriminant of an imaginary quadratic field of class number divisible by 5, so that the corresponding subfield of the Hilbert class field gave a D_5 extension of $\mathbb{Q}$.

If the above test failed (i.e. no small prime factored as a 3-cycle) then the following definitive criterion was applied: The galois group of an irreducible quintic with square discriminant is a proper subgroup of A_5 if and only if the sextic resolvent of the quintic has a rational root. The sextic resolvent of a quintic is described in appendix 1; in a sense that will be made precise in the next chapter, the sextic provides more information about the A_5 field than the quintic. At any rate, a small number of quintics did not factor as a 3-cycle modulo any prime less than 100 and yet the corresponding sextics did not have rational roots. Further search for larger 3-cycle primes was then always successful.

5. The only serious technical problem in the above procedure was the determina-

tion of the ring of integers in the root field of a quintic polynomial. The integral closure algorithm of Zassenhaus (see [Zassenhas] or [Zimmer]) was modified to give an efficient technique for solving this problem. A variety of special methods can be used for specific polynomials, but an algorithmic procedure is necessary if the computations are to be done by a machine. Note that in Table 3.3 the six lowest conductors are associated to quintics with spurious factors in their discriminant; it seems likely that none of them have power bases.

6. A cursory inspection of Table 3.1 reveals that the knowledge of the discriminant together with the factorization of the prime is sufficient to determine the "type" of the prime, with one exception. The types 7 and 9 give the same discriminant and factorization. Appendix 2 gives a simple criterion for discriminating between these possibilities.

7. A quintic with square (and hence positive) discriminant must have either 1 or 5 real roots. One expects those quintics with small discriminants to have 1 real root. And if one is interested in finding modular forms via odd two dimensional representations then the totally real quintics must be ruled out. The totally real fields in the Table A5 are indicated by an asterisk; as expected there are only a few (4) of them and they have larger conductors.

8. The field of conductor 800 is a remarkable concatenation of interesting peculiarities:

a) it has the lowest conductor of any field in the search (compare with the conductors for the known dihedral, tetrahedral and octahedral representations described in [Serre, D]).

b) the lemma of appendix 2 alluded to above is necessary,

c) at 2 the local extension is the unique A_4 extension of $\mathbb{Q}_2$ and hence the only possible primitive local representation that can occur as a local component of an icosahedral representation,

d) at 5 the representation corresponds to a wildly ramified cyclic extension of degree 5 that is the only possible local representation for which the distinction between the two projective representations of A_5 is troublesome. As we will see in the subsequent chapters this ambiguity at 5 makes it difficult to compute the fifth coefficient of the L-series of the representation of conductor 800.

Table 3.1: Ramified Primes in A_5 Fields

type	ramification groups	splitting	discr.	cond	p
1	C_5, C_5	1^5	4	1	$p \equiv 1(5)$
2	C_3, C_3	$1^3 1 1$	2	1	$p \equiv 1(3)$
3	C_2, C_2	$1^2 1^2 1$	2	1	$p \neq 2$
4	D_5, C_5	1^5	4	2	$p \equiv -1(5)$
5	D_3, C_3	$1^3 2$	2	2	$p \equiv -1(3)$
6	D_2, C_2	$2^2 1$	2	2	$p \neq 2$
7	C_5, C_5, C_5	1^5	8	2	$p = 5$
8	D_5, D_5, C_5	1^5	6	3	$p = 5$
9	D_5, C_5, C_5	1^5	8	4	$p = 5$
10	C_3, C_3, C_3	$1^3 1 1$	4	2	$p = 3$
11	D_3, D_3, C_3	$1^3 1^2$	4	3	$p = 3$
12	D_3, C_3, C_3	$1^3 2$	4	4	$p = 3$
13	D_3, D_3, C_3, C_3, C_3	$1^3 1^2$	6	5	$p = 3$
14	C_2, C_2, C_2	$1^2 1^2 1$	4	2	$p = 2$
15	C_2, C_2, C_2, C_2	$1^2 1^2 1$	6	3	$p = 2$
16	D_2, C_2, C_2	2^2	4	4	$p = 2$
17	A_4, D_2, D_2	$1^4 1$	6	5	$p = 2$
18	D_2, C_2, C_2, C_2	$2^2 1$	6	6	$p = 2$
19	D_2, D_2, D_2, C_2, C_2	$1^4 1$	8	7	$p = 2$

Table 3.2: Conductors of Quintic Fields

The notation in the table below is as follows. Let $f(x) = x^5+a_1x^4+a_2x^3+a_3x^2+a_4x+a_5$ be a quintic polynomial with integral coefficients and square discriminant $= d^2$. Assume that the discriminant of a root field of f(x) is D^2, that f(x) is irreducible mod q_1 and factors as a cubic times two distinct linear factors mod q_2. Finally let C be the minimal conductor of a lifting of a projective representation corresponding to f(x), and for each prime p_i dividing the discriminant of f(x) let t_i be the "type" of its ramification in terms of Table 3.1. Then the entry associated to f(x) in the above table is written:

C = factorization of C $[a_1,a_2,a_3,a_4,a_5](q_1,q_2;D:d)$ $p_1[t_1]$, $p_2[t_2]$ ...

If the type is 20 (resp. 21) this means type 2 or type 3 (resp. type 5 or type 6). The type "un." means that the prime is unramified in a root field of f(x) i.e. that the prime divides d but not D. This table is continued in Table A_5 at the end of this thesis; in that table the totally real fields are marked with an asterisk.

$800 = 2^5 5^2$	[0, 10,-10, 35,-18]	(7, 3; 55000: 5000)	2[17], 5[7], 11[un.]
$837 = 3^3 31^1$	[4, 25, 20, 8, 5]	(2, 5; 77841: 8649)	3[11], 31[1]
$992 = 2^5 31^1$	[9, 20, 36,-11, 1]	(5, 7; 46128: 7688)	2[17], 3[un.], 31[1]
$1161 = 3^3 43^1$	[4, 25, 17, 5, 2]	(5, 2; 31347: 387)	3[11], 43[2]
$1188 = 2^2 3^3 11^1$	[4, 13, -5, -2, 1]	(5, 7; 4356: 2178)	2[5], 3[11], 11[1]
$1376 = 2^5 43^1$	[2, 6, 8, 10, 8]	(3, 7; 688: 344)	2[17], 43[2]
$1501 = 19^1 79^1$	[6, 19, 25, 11, 2]	(3, 2; 1501: 1501)	19[3], 79[2]
$1600 = 2^6 5^2$	[5, 20, 0, 15, -1]	(3, 7; 80000: 5000)	2[18], 5[7]
$1687 = 7^1 241^1$	[2, 3, 2, 12,-16]	(3,17; 26992: 1687)	2[un.], 7[3], 241[2]
$1825 = 5^2 73^1$	[4, -1,-21, -1, -7]	(2, 7; 45625: 365)	5[21], 73[2]
$1948 = 2^2 487^1$	[0, 20,-20,-16,-16]	(3, 7; 498688: 974)	2[5], 487[3]
$2083 = 2083^1$	[1, 5,-11, 4, 1]	(2,11; 2083: 2083)	2083[3]

Table 3.3: Primitive Extensions of $\mathbb{Q}_2$

f(x)	resolvent	discr_f	G	r	$A(T_2)$	(T,2)
$x^4+2x^3-2x^2+2$	$y^3+2y^2-8y-24$	$-5568 = 2^6(-87)$	A_4	1	5	-1
x^4-2x+2	y^3-8y-4	$1616 = 2^4(101)$	S_4	1	3	1
x^4-4x+2	$y^3-8y-16$	$-4864 = 2^8(-19)$	S_4	5	7	1
x^4-4x^2+4x-2	$y^3+4y^2+8y+16$	$-2816 = 2^8(-11)$	S_4	5	7	-1

The splitting fields over $\mathbb{Q}_2$ of the above four polyomials give the four primitive extensions of $\mathbb{Q}_2$; three with galois group $G = S_4$ and one with galois group A_4. It is easy to verify that the galois groups are as claimed by using the discriminants and cubic resolvents. The cubic extension of $\mathbb{Q}_2$ in the A_4 extension is the unramified extension of $\mathbb{Q}_2$ of degree 3. The S_3 extension of $\mathbb{Q}_2$ in the S_4 extensions is the unique S_3 extension of $\mathbb{Q}_2$ (which has ramification index 3 and residue class degree 2). This information enables one to easily write a formula for $r = b_0(K/\mathbb{Q}_2)$ in terms of the 2-valuation of the discriminant; e.g. in the S_4 cases, $3+r = v_2(\text{discriminant})$. The minimal conductor is computed by using Theorem 4, and (T,2) is computed by using the global S_4 or A_4 extensions determined by the above polynomials and then using the fact that $\prod(T,p) = 1$ together with the information given in the text on (T,p) for non-primitive T_p. The assertion that $(T,2) = 1$ is equivalent to T_2 having a lifting of index 2 so that these quantities can be computed purely locally (with some difficulty) by using the ideas mentioned at the end of chapter 1 (following [Weil, ED]).

CHAPTER FOUR

Let K be a galois extension of $\mathbb{Q}$ and let $T:G(K/\mathbb{Q}) \longrightarrow PGL_2(\mathbb{C})$ be a faithful representation. If $\chi:C_K \longrightarrow \mathbb{C}^*$ is centric for T then T and χ determine a lifting

$$\mathbb{T}:G_{\mathbb{Q}} \longrightarrow GL_2(\mathbb{C})$$

up to twisting by elements of the character group $\mathrm{Hom}(G(K/\mathbb{Q}),\mathbb{C}^*)$.

For the remainder of this chapter let T be an icosahedral representation so that K is an A_5 extension of $\mathbb{Q}$. Since A_5 is simple and hence has no nontrivial characters, we see that in this case the projective representation T together with a centric character χ uniquely determine a lifting $\mathbb{T}$. In particular if one "knows" T and χ it should be possible to compute the coefficients of the L-series

$$L(s,\mathbb{T}) = \Sigma a_n n^{-s}.$$

The object of this chapter is to describe this computation; later on we will take K to be the specific field mentioned at the end of the last chapter. With slight modifications all of the techniques apply to any A_5 field; with further modifications they apply to any two-dimensional representation.

A character $\chi:C_K \longrightarrow \mathbb{C}^*$ is centric for T if it induces the nontrivial character on the group

$$H^{-1}(G(K/\mathbb{Q}),C_K) \cong C_K{}^N/C_K{}^I$$

of order two. Hence any odd power of a given χ (including χ^{-1}) is also centric for T and so it is always possible to take χ to be of 2-power order. Also note that χ^2 is trivial on $C_K{}^N$ so that χ^2 is the composition of a character of $C_{\mathbb{Q}}$ with the norm map $N_{K/\mathbb{Q}}$. By the definition of the centric character of $\mathbb{T}$ it is clear that $\chi^2 = \det(\mathbb{T})\circ N_{K/\mathbb{Q}}$.

For each rational prime p choose a prime P of K lying over p. Let χ be centric for T so that in particular χ is invariant under the action of $G(K/\mathbb{Q})$. This implies that the component of χ at P determines the components of χ at the other primes of K above p. It is always possible to choose χ to be unramified at primes above rational primes that are unramified in $K/\mathbb{Q}$; we will always assume that χ has been so chosen. If P is unramified then we will follow the usual abuse of notation by letting $\chi(P)$ denote the value of χ on an idele that is 1 outside P and has valuation 1 at P.

Let L/K be the cyclic extension of K corresponding to χ, so that the corresponding lifting $\mathbb{T}$ induces a faithful representation $G(L/\mathbb{Q}) \longrightarrow GL_2(\mathbb{C})$. Assume that p is unramified in $K/\mathbb{Q}$ (and hence in $L/\mathbb{Q}$) and let $F_p \in G(L/\mathbb{Q})$ be a frobenius element. Let

$$t_p = \mathrm{trace}(\mathbb{T}(F_p)) \qquad d_p = \mathrm{determinant}(\mathbb{T}(F_p))$$

so that the Euler factor associated to p in $L(s,\mathbb{T})$ is

$$(1-t_p p^{-s}+d_p p^{-2s})^{-1}.$$

Note that t_p^2/d_p depends only on the image of $\mathbb{T}(F_p)$ in $PGL_2(\mathbb{C})$, i.e. on $T(F_p)$ (this is a slight abuse of notation since $F_p \in G(L/\mathbb{Q})$ but the image of F_p in $G(K/\mathbb{Q})$ is a frobenius element for p in $G(K/\mathbb{Q})$ ([Serre, CL, p.34]) so this is harmless.)

Now we want to show how t_p can be calculated from information derived from χ, T, and $\det(\mathbb{T})$. This computation depends upon the order of the image of F_p in $G(K/\mathbb{Q})$. Call an unramified prime an "n-prime" if the residue class degree of P is n, so that the order of F_p in $G(K/\mathbb{Q})$ is n (and here n= 1,2,3, or 5). Another way to express this condition is that n is the least integer such that $F_p^n \in G(L/K)$. By a basic property of the frobenius substitution ([Serre, CL, p.23]) F_p^n is exactly the frobenius substitution in G(L/K) of a prime P in K lying above p. This implies that $\mathbb{T}(F_p)^n$ is a scalar matrix whose entry on the diagonal is $\chi(P)$.

Proposition: Let p be an n-prime, let $t = t_p = \text{trace}(\mathbb{T}(F_p))$, $d = d_p = \det(\mathbb{T}(F_p))$, and $x = \chi(P)$. Then for n= 1,2,3,5 we have the following relationships between t,x and d:

n= 1: $t = 2x$

n= 2: $t = 0$

n= 3: $t = -x/d$

n= 5: $t = x/(d^2(2-u))$ where u satisfies $u^2-3u+1 = 0$.

Remarks: The point is that t is determined by x, d, and t^2/d which depend upon (respectively) X, $\det(\mathbb{T})$ and T. There is no similar result for n= 4 which is consistent with the fact that X and T do not uniquely determine a lifting if T is octahedral.

Proof: Let $X = \mathbb{T}(F_p)$ so that X satisfies the Cayley-Hamilton equation $X^2-tX+dI = 0$ (where I is the identity matrix). Then:

n = 1: Here X is already a scalar matrix so that $t = 2x$.

n = 2: Since X^2 is a scalar matrix, X is not a scalar matrix, and $X^2-tX+dI = 0$, we must have $t = 0$.

n = 3: The equation $X^2-tX+dI = 0$ implies that $X^3 = (t^2-d)X-dtI$. Thus $t^2 = d$ and $t = -x/d$.

n = 5: Here $X^5 = (t^4-3t^2d+d^2)X+(2dt-dt^3)I$ which means that $u = t^2/d$ must satisfy the equation $u^2-3u+1 = 0$ and that $t = \chi(P)/(d^2(2-u))$.

Thus the calculation of the Euler factors of $L(s,\mathbb{T})$ at unramified primes is easily reduced to the determination of $\chi(P)$ and $T(F_p)$. The ramified primes will be handled later.

The Calculation of T

If we explicitly realize $G(K/\mathbb{Q})$ as the group of symmetries of an icosahedron via T then it is easy to determine the conjugacy classes in $G(K/\mathbb{Q}) \cong A_5$. Rotations about the same angle are conjugate. A rotation of order 2 interchanges two of the antipodal vertices of the icosahedron so that a rotation through an angle θ is conjugate to a rotation through an angle $-\theta$. Thus the conjugacy classes of the group are the conjugacy class containing only the trivial rotation, the conjugacy class of rotations through an angle π, the conjugacy class of rotations through an angle $\pm 2\pi/3$, the conjugacy class of rotations through an angle $\pm 2\pi/5$, and the conjugacy class of rotations through an angle $\pm 4\pi/5$.

The icosahedron has 30 edges, and there are 15 diagonals that connect pairs of midpoints of these edges. Call three of these diagonals a "triad" if they are mutually perpendicular. Then these 15 diagonals determine 5 triads and the group of symmetries of the icosahedron induces all even permutations of these 5 triads. This gives an intrinsic realization of the icosahedral group as the group of even permutations on 5 objects.

The representation T is determined if one knows the conjugacy class of $T(F_p)$ for almost all unramified primes p. One way to describe this information is to give t_p^2/d_p for each unramified prime p. By the computations in the claim above we have:

class of F_p	—>	t_p^2/d_p
1-prime	—>	4
2-prime	—>	0
3-prime	—>	1
one class of 5-primes	—>	u
the other class of 5-primes	—>	u'

where u and u' are the distinct roots of $x^2-3x+1=0$. The choice of specific u and u' specifies one or the other of the two possible faithful projective representations of $G(K/\mathbb{Q})$.

If $f(x)$ is a quintic polynomial whose splitting field is K then the factorization of $f(x)$ in $\mathbb{Q}_p$ exactly mirrors the cycle structure of F_p in $G(K/\mathbb{Q}) \cong A_5$. This determines the conjugacy class of F_p except when F_p is a 5 cycle.

There are two conjugacy classes of 5-cycles in $G(K/\mathbb{Q})$ and a nice idea due to Serre can be used to distinguish between them. Suppose that $f(x)$ is irreducible over $\mathbb{Q}_p$ so that F_p is a 5-cycle in $G(K/\mathbb{Q})$. Let $\mathrm{discr}_f = D^2$ and let y be a root of $f(x)= 0$ in the unramified extension of degree 5 of $\mathbb{Q}_p$. Then

$$\prod_{0\leq i<j\leq 4} (F_p^{\,i}(y)-F_p^{\,j}(y)) = \pm D \quad \text{in } \mathbb{Q}_p$$

and the $\pm$ distinguishes between the two conjugacy classes of 5-cycles in $G(K/\mathbb{Q})$.

For the remainder of this chapter let K be the splitting field of the quintic

$$f(x)= x^5+10x^3-10x^2+35x-18.$$

The discriminant of $f(x)$ is $\mathrm{discr}_f = 2^6 5^8 11^2$, and the discriminant of a root field of $f(x)$ is $2^6 5^8$. The implementation of the above prescription for the computation of the conjugacy classes of the F_p ($p \neq 2,5$) is straightforward. The cycle structure is determined by factoring in $\mathbb{Q}_p$ but of course if $p \neq 11$ then this factorization is determined by the factorization mod p. In fact in the A_5 case the cycle structure is determined by the number of roots mod p which is equal to the gcd of $f(x)$ and x^p-x. The resolution of the ambiguity at the 5-cycles can be computed by computing the above product modulo p in which case F_p is just the p-th power mapping. The results of these computations are exhibited in table 4.1.

The mapping

$$p \longrightarrow t_p^2/d_p, \qquad p \text{ unramified}$$

determines T. However it is not completely clear as to how this information determines the local projective representations at the ramified primes. Let I_5 be the inertia group (which is also the decomposition group) at a prime above 5 in $G(K/\mathbb{Q})$. Then I_5 is cyclic of order 5, and there are two faithful two-dimensional projective representations of a cyclic group of order 5 (if λ is a faithful character of the group of order 5 then the projectivizations of $\lambda \oplus \lambda^{-1}$ and $\lambda^2 \oplus \lambda^{-2}$ are distinct projective representations). These two projective representations are interchanged by reversing u and u' above, but later we will have to grapple with the question of which local representation is induced by a given choice of u and u' in order to determine the Euler factor of $L(s,\mathbb{T})$ at 5.

The Computation of χ

According to theorem 5 we can uniquely specify a lifting of T by giving local liftings (almost everywhere unramified) for the $T_p = T|G(K_p/\mathbb{Q}_p)$. The global lifting $\mathbb{T}$ then agrees with the local liftings on the inertia groups. This procedure also fixes χ and $\det(\mathbb{T})$.

For our specific K we can choose the local liftings to be unramified outside 2 and 5. At 2 we choose a lifting of minimal conductor and at 5 we take a lifting of index 1 and determinant 1. At p=5 the representation then looks like $\lambda \oplus \lambda^{-1}$, where λ determines the extension of $K_p/\mathbb{Q}_p$. We can think of λ as a character on $\mathbb{Q}_5^*$ that is of order 5 and conductor 5^2. If $\mathbb{T}$ is the global lifting determined by these local liftings then the centric character χ of $\mathbb{T}$ is unramified outside 2 and the component of χ on the units of the ring of integers of K_2 can be taken to be the character described in appendix 3. The determinant of $\mathbb{T}$ is unramified outside 2 and by the computations in appendix 3 it is equal to the unique Dirichlet character μ of order 2 and conductor 2^2. The conductor of $\mathbb{T}$ is $2^5 5^4 = 20000$. After computing the $\chi(P)$ to obtain $L(s,\mathbb{T})$ we will twist $\mathbb{T}$ so that the local representation at 5 is $1 \oplus \lambda^{-2}$ which then gives a conductor of $2^5 5^2 = 800$.

At first glance the computation of χ appears to be out of the realm of possibility since it is defined on the group of idele classes of an extension of $\mathbb{Q}$ of degree 60. However if a subgroup $G(K/E)$ of $G(K/\mathbb{Q})$ has trivial Schur multiplier then Theorem 2 says that there is a character $\psi: C_E \longrightarrow \mathbb{C}^*$ with $\chi = \psi \circ N_{K/E}$. The largest subgroup of A_5 with trivial multiplier is the dihedral group of order 10, D_5. Let E be the field fixed by such a subgroup of $G(K/\mathbb{Q})$. Then

$$H^2(G(K/E),\mathbb{C}^*)= \{0\}.$$

This marvelous fact reduces the computation of χ to computations in the sextic field E.

The formulae in appendix 1 can be applied to the specific quintic above to yield a polynomial

$$g(x)= x^6-1000x^4+240000x^2-1760000x+8000000$$

whose root field isomorphic to E. The discriminant of $g(x)$ is $\mathrm{discr}_g = 2^{48}3^6 5^{28}109^2 \cong 10^{41}$. If y is a root of $g(x)$ it is more convenient to use the polynomial for $y^2/40$ as a model for E. This polynomial is

$$h(x) = x^6-50x^5+925x^4-7250x^3+16250x^2+7250x+15625$$

which has $\mathrm{discr}_h= 2^6 3^6 5^{22}11^6 109^2 \cong 10^{31}$. The decomposition of the various rational primes is described in table 4.2.

Any two characters ψ that satisfy $\chi = \psi \circ N_{K/E}$ must have a quotient that is trivial on $N_{K/E}C_K$. Let E' be the fixed field of the subgroup of order 5 of $G(K/E)$. Then $G(E'/E)$ is the maximal abelian quotient of $G(K/E)$ so

$$N_{K/E}C_K= N_{E'/E}C_{E'}.$$

Let $\Theta: C_E \longrightarrow \pm 1$ correspond to the nontrivial character of $G(E'/E)$. Then there are precisely two $\psi: C_E \longrightarrow \mathbb{C}^*$ such that $\chi = \psi \circ N_{K/E}$, and their quotient is Θ.

Note that $\chi^2 = \mu \circ N_{K/\mathbb{Q}}$ so that χ is of order 4. If L is (as above) the extension of K determined by χ then $G(L/E)$ is isomorphic to the product of a cyclic group of order 4 and D_5. Since the largest cyclic quotient of this group is of order 4 it is clear that ψ is also of order 4.

In the field E there is a unique prime P_2 above 2; P_2 is of residue class degree 3 and ramification index 2 over $\mathbb{Q}$. The completion E_2 of E at P_2 is isomorphic to a sextic extension of $\mathbb{Q}_2$ contained in the unique A_4 extension of $\mathbb{Q}_2$.

The rational prime 5 splits in E as

$$(5) = P_5 (P_5')^5$$

and K/E is unramified outside P_2 and P_5. Since χ is unramified outside of the primes of K above 2 it is clear that ψ is unramified outside of P_2 and P_5. The local norm index of K/E at P_5 is 5 and is 1 at P_5'. Since we have chosen χ to be of 2-power order it follows that that ψ is unramified at the finite places outside P_2.

At P_2, ψ can be taken to be the character described in appendix 3. By multiplying by Θ_2 or by changing the sign of i we can get 4 different characters on the units at 2; fixing one of these has the effect of definitively fixing the character χ.

There are two real archimedean places of E and two complex places. The character Θ is ramified at both of the real places. The results in appendix 3 show that $\Theta_2(-1) = -1$. By looking at the principal idele -1 it is easy to see that must be ramified at precisely one real place. Combining the above remarks yields:

<u>Proposition: There is a character</u> $\psi: C_E \longrightarrow \mathbb{C}^*$ <u>such that:</u>

a) $\psi \circ N_{K/E}$ <u>is centric for</u> T

b) ψ is of order 4

c) ψ is unramified outside P_2 and the real archimedean places

d) ψ is given on the units at P_2 by the formula in appendix 3

e) ψ is ramified at precisely one of the two real archimedean places.

Certainly there are only a finite number of ψ satisfying the last four conditions; the number of possibilities depends roughly on the size (and structure) of the ideal class group of E. However there is one more property of ψ that serves to specify it uniquely and makes the computations drastically easier.

This additional constraint is that $\psi \circ N_{K/E}$ is invariant under $G(K/\mathbf{Q})$. In order to explore this condition it is convenient to examine separately each of the 4 types of unramified primes. It will turn out that the galois invariance of will give us semi-local conditions on ψ; namely, relationships between the values of ψ on the various primes of E over a given odd rational prime p.

Suppose that p is a 3-prime that splits as (p)= PP' in E, where P and P' are primes of E that have residue class degree 3 over $\mathbf{Q}$. Then P and P' split completely in K/E so that $\Theta(P)= \Theta(P')= 1$. In fact P and P' are norms of primes of K that are conjugate and consequently $\psi(P)= \psi(P')$. Let $x(p) \in J_E$ be the principal idele corresponding to $p \in E^*$. Then

$$x(p) = x_2 x_P x_{P'} u$$

where the subscripts indicate the components of x(p) at 2, P, and P' and u is an idele that is positive at the real places and a unit outside 2,P and P' so that $\psi(u) = 1$. If v is the uniformizing parameter in E_2 as described in appendix 3 then

$$x_2 = 1 \text{ or } 1+v^2 \bmod (\text{kernel } \psi_2)$$

depending on whether $p \equiv 1$ or $3 \bmod 4$. Using appendix 3 it is easy to work out that $\psi(x_2) = \mu(p)$. Thus

$$1 = \mu(p)\,\psi(P)\,\psi(P')$$

which says that $\psi(P)^2 = \mu(p)$. Thus we know $\psi(P)$ up to a sign.

Now suppose that p is a 5-prime that splits as $(p) = PQ$ where P is of degree 1 and Q is of degree 5. If $\overline{Q}$ is a prime of K above Q then some conjugate of $\overline{Q}$ has norm P^5. This implies that

$$\psi(P)^5 = \psi(P) = \psi(Q)$$

(ψ is of order 4). Again we consider the principal idele p and conclude that

$$1 = \mu(p)\;\psi(P)\;\psi(Q)$$

which means that again ψ is determined on the primes above p by giving a specific square root of $\mu(p)$.

Suppose that p is a 1-prime that splits as

$$(p) = P_1P_2P_3P_4P_5P_6.$$

By the arguments used above

$$\psi(P_i) = \psi(P_j) \text{ for all } i,j$$

and

$$\psi(P_i)^6 = \psi(P_i)^2 = \mu(p).$$

Suppose that p is a 2-prime so that it splits as

$$(p) = PP'QQ'$$

where P and P' are of degree 1 and Q and Q' are of degree 2. The arguments used above imply that

$$\psi(P)^2 = \psi(P')^2 = \psi(Q) = \psi(Q')$$

and $1 = \mu(p)\,\psi(P)\,\psi(P')\,\psi(Q)\,\psi(Q')$. In this case we need some more information in order to reduce the computation of ψ to a question of sign.

Lemma: *With the immediately preceding notation we have*

$$\psi(P) = -\psi(P').$$

Remark: Granting the lemma, the formulae preceding the lemma show that $\psi(P)^2 = -\mu(p)$ and that once again the values of ψ on all of the primes above p are completely specified by giving the choice of a square root (in this case the square root of $-\mu(p)$). The calculations of ψ^2 can be summarized by saying that $\psi^2 = \mu\,\Theta$.

Proof: Let $X = \mathbb{T}(F_p) \in GL_2(\mathbf{C})$. Then X^2 is a diagonal matrix with $\chi(\overline{Q})$ on the diagonal, where $\overline{Q}$ is a prime of K above Q. The trace of X is 0 and the Cayley-Hamilton equation

$$X^2 + \det(X)I = 0$$

says that $\chi(\overline{Q}) = -\mu(p)$. Thus

$$-\mu(p) = \chi(\overline{Q}) = \psi(Q) = \psi(Q') = \psi(P')^2 = \psi(P)^2$$

and if this information is used in the equation

$$1 = \mu(p)\ \psi(P)\ \psi(P')\ \psi(Q)\ \psi(Q')$$

the formula in the statement of the lemma results. q.e.d.

If the above calculations imply that if $x \in J_E$ is a unit at all 2-primes then $\psi(x)$ depends only on the component of x at 2, the sign of x at the real places, and the norm of x. More generally, if we fix a specific prime P of degree one above each 2-prime p then $\psi(x)$ for any idele x depends on the above information together with the parity of $v_P(x)$ above each 2-prime p.

It is now possible to compute the values of ψ by using relations generated by looking at principal ideles. The results of a computer program that "computed" ψ are tabulated in Tables 4 3, 4.4 and 4.5.

Table 4.3 is a collection of principal ideles. The notation is as follows. In a given line of the table, the principal idele whose norm is being considered is given at the left by giving its coefficients as a polynomial in a root of the polynomial h given above (constant coefficient last). Call the principal idele x. The 4-tuple in brackets contains 1) the sign of x at one of the real places, 2) the sign of x at the other real place, 3) the value of Θ_2 on the component of x at 2, 4) the value of the character ψ_2 of appendix 3 on x multiplied by $(-1)^s$ where s is the number of 2-primes p for which $v_P(x)$ is odd. In all cases the notation is "additive" so that, for instance, 0 (resp. 1) in the first component means that the idele is positive (resp. negative) at the first real place and if there is a 1 (resp. 3) in the fourth component then $\psi_2(x_2) = i$ (resp. $-i$) if there are no 2-primes in x. Finally the rest of the line associated to x contains the factorization of the norm to $\mathbb{Q}$ of x; this is represented in the form of a string of expressions p:n (which means that the p-part of the norm is p^n).

Ordinarily the computation of an idele class character requires such a table as 4.3 except that it is necessary to know the factorizations of principal ideles in the field E. Here the results given above almost determine the character by knowing the factorizations of the norms of the principal ideles; the only further information required is the parity of the various two-primes in the principal ideles x.

Table 4.3 can be viewed as containing "relations" on the values of ψ at the various primes in E. For instance, the first relation says that

$$1 = i \ \psi(P_3)^3 \psi(P_5)^4 \psi(P_{11})^2.$$

These relations can be "reduced" to determine the values of ψ (modulo the fact that we don't yet know which real prime ψ is ramified at). For instance, since we know ψ^2, the first relation determines the value of ψ on either prime of E above 3. The algorithm used to systematically perform this reduction process involves an interesting application of some techniques for representing and manipulating sparse matrices. The results are contained in Table 4.4. In this reduction process units sometimes turn up; in the notation of Table 4.1 they are

[0,0,0,0] [1,1,0,2] [1,0,1,0] [0,1,1,2].

The first two units are expected; they correspond to ± 1. The third entry is nontrivial and shows that ψ is ramified at the second archimedean prime (which corresponds to the largest real root of h(x)), and is unramified at the other real place.

Note that there is no asymmetry here because we have fixed our choice of ψ_2. If ψ were multiplied by Θ then it would become ramified at the other real place.

The values of the L-series can be readily computed from Tables 4.1, 4.4 and the rules described above. The results are contained in Table 4.5. The coefficient of 5 is 0 since the restriction of T to a decomposition group at 5 does not involve the identity character. Therefore any coefficient not relatively prime to 2 and 5 must be 0.

The limiting factor on extending these calculations further is the problem of calculating the values of ψ at 3-primes. It is hard to find principal ideles whose norms involve only primes less than 360 and yet involve a given 3-prime (which means that the principal idele must be a cubic polynomial in the root of h(x) and which means that the norm must be divisible by the cube of the prime). To find the values of ψ on the last few 3-primes in the table it was necessary to write a special program to search for appropriate principal ideles. The amount of computation required seemed to go up exponentially so that it would be hard to extend Table 4.4 much further (at the 3-primes). It is fortunate that we will need only (roughly) the first 360 coefficients of $L(s,\mathbb{T})$.

This difficulty at the 3-primes seems only fair; the 2-primes are distinguished by the difficulties described above, the 5-primes are distinguished by the ambiguities discussed below, and the 1-primes (i.e. primes that split completely) are distinguished by their absence in the first 200 primes.

The Ramified Primes

Let $\mathbb{T}$ be the lifting of T determined by the χ computed above, and let $L(\mathbb{T},s) = \Sigma a_n n^{-s}$ be its L-series. Since the restriction of $\mathbb{T}_2$ to the inertia group is irreducible we must have $a_2 = 0$. The restriction of $\mathbb{T}_5$ to the inertia group is reducible but it does not involve the trivial character so the fifth coefficient a_5 of $L(\mathbb{T},s)$ is also 0.

However $\mathbb{T}$ is of conductor $2^5 5^4$, and to get a representation, say $\mathbb{T}'$, of conductor $2^5 5^2 = 800$ it is necessary to twist by an appropriate Dirichlet character of conductor 25 and order 5. There are four such characters; if twisting by λ gives a representation of conductor 800 then so does twisting by λ^{-1} but twisting by λ^2 or λ^{-2} still gives a representation of conductor 20000. The following lemma determines which is the correct character to twist by. Note that any character $\lambda':\mathbf{Q}_5^* \longrightarrow \mathbf{C}^*$ determines (essentially by restriction to the units) a Dirichlet character whose conductor is a power of 5. Recall that E' is an extension of degree 12 of $\mathbb{Q}$ contained inside K.

Proposition: Let p be a 5-prime; choose $F_p \in G(K/\mathbb{Q})$ so that F_p lies in $G(K/E')$. Choose P in E' lying over p such that the frobenius of P in $G(K/E')$ is F_p (P is one of the two primes of E' of degree 1 over p). Assume that the projective representation T has been fixed so that $t_p^2/d_p = tr(F_p)^2/det(F_p) = u$ where u is a specific root of $x^2-3x+1 = 0$. Let $\lambda_{K/E'}: C_{E'} \longrightarrow \mathbb{C}^*$ be a character that determines K/E' and that has the property that

$$2 + \lambda_{K/E'}(P) + \lambda_{K/E'}(P)^{-1} = u.$$

Then the local component of $\lambda_{K/E'}$ at one of the primes of E' above 5 of degree 1 is a character $\lambda':\mathbb{Q}_5^* \longrightarrow \mathbb{C}^*$ such that $\mathrm{cond}(\mathbb{T}\otimes\lambda') = 800$.

The proof is a straightforward exercise in algebraic number theory; it involves showing that the projective representation of $I_5 \cong G(K/E')$ determined by T is the projectivization of $\lambda \oplus \lambda^{-1}$. Unfortunately the 12-th degree field E' is very difficult to compute with; it is even difficult to produce a polynomial whose root field is E'. Thus the above lemma is not practical. However in the last chapter it will be seen that it is possible to find the correct λ "almost certainly" by using the fact that $\mathbb{T}'$ is supposed to correspond to a modular form of weight one.

Table 4.1

The following is a tabulation of the conjugacy classes of the frobenii in the A_5 field given in the text. The conjugacy class of elements of order 2 is denoted "2-2", the conjugacy class of 3-cycles is denoted "3", and the two conjugacy classes of 5-cycles are denoted "5A" and "5B".

3: 3	7: 5A	11: 2-2	13: 5A
17: 2-2	19: 5B	23: 3	29: 5B
31: 3	37: 3	41: 2-2	43: 5B
47: 5A	53: 3	59: 5B	61: 3
67: 2-2	71: 5A	73: 3	79: 5B
83: 3	89: 2-2	97: 5A	101: 3
103: 5B	107: 3	109: 2-2	113: 5B
127: 2-2	131: 5A	137: 3	139: 3
149: 3	151: 5B	157: 3	163: 3
167: 5B	173: 3	179: 3	181: 2-2
191: 2-2	193: 2-2	197: 2-2	199: 5A
211: 3	223: 2-2	227: 2-2	229: 2-2
233: 2-2	239: 3	241: 3	251: 3
257: 5B	263: 5A	269: 2-2	271: 2-2
277: 3	281: 5B	283: 5A	293: 2-2
307: 2-2	311: 5A	313: 2-2	317: 3
331: 2-2	337: 5A	347: 5B	349: 2-2
353: 5A	359: 5A	367: 5A	373: 5B
379: 3	383: 3	389: 3	397: 5B
401: 3	409: 3	419: 3	421: 2-2
431: 3	433: 5B	439: 5A	443: 3
449: 5B	457: 2-2	461: 3	463: 3
467: 5A	479: 5B	487: 3	491: 3
499: 5A	503: 2-2	509: 5A	521: 3
523: 5B	541: 2-2	547: 3	557: 2-2
563: 5A	569: 3	571: 3	577: 2-2
587: 3	593: 5A	599: 3	601: 5A
607: 5A	613: 5A	617: 3	619: 3
631: 3	641: 5A	643: 2-2	647: 2-2
653: 5A	659: 2-2	661: 5B	673: 2-2
677: 5A	683: 3	691: 5B	701: 2-2

The following is the set of primes less than 32768 that split completely:

2063, 2213, 2953, 3631, 3643, 4139, 5519, 5689, 6833,
6971, 7499, 7907, 8623, 9311, 10333, 10799, 12979, 13241, 15017,
15199, 15227, 16057, 16481, 17417, 17923, 18637, 19069, 19687, 19699,
19709, 19751, 21341, 21673, 21859, 21937, 21997, 22343, 23561, 24229,
24251, 25423, 27397, 27581, 27767, 27983, 28393, 29851, 31769, 32059

Table 4.2: Primes in K and its subfields

The following table describes the behavior of primes in the A_5 field given in the text. The description of the behavior of the unramified primes is of course valid for any A_5 field. The notation is as in the text with the additional notation that $\Sigma n_i(f_i^{e_i})$ means the product over i of n_i primes of residue class degree f_i and ramification index e_i. The field F is a root field of the quintic polynomial whose splitting field is K.

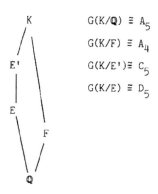

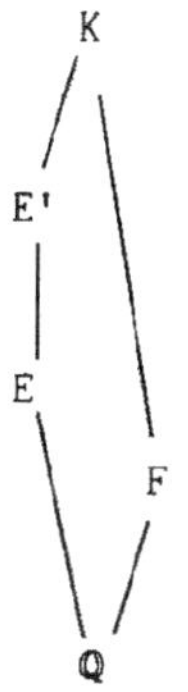

$G(K/\mathbb{Q}) \cong A_5$

$G(K/F) \cong A_4$

$G(K/E') \cong C_5$

$G(K/E) \cong D_5$

field	1-primes	2-primes	3-primes	5-primes	2	5
K	60(1)	30(2)	20(3)	12(5)	$5(3^4)$	$12(1^5)$
E'	12(1)	6(2)	4(3)	1+1+5+5	3^4	1^5+1^5+1+1
E	6(1)	1+1+2+2	3+3	1+5	3^2	1^5+1
F	5(1)	1+2+2	1+1+3	5	1^4+1	1^5

Table 4.3: Some Norms

The notation is explained in the text.

0,0,0,0,1,5:	[0,0,0,3] ;	3:3, 5:4, 11:2,
0,0,1,-6:	[0,1,0,0] ;	7:2, 11:3,
0,-1,5,25:	[1,0,1,3] ;	5:8, 19:1, 109:2,
0,0,-1,10:	[1,0,0,2] ;	3:3, 5:4, 17:1,
0,-1,16,60:	[0,0,0,0] ;	3:3, 5:4, 17:2, 109:1, 127:1,
0,-1,15,20:	[1,0,0,1] ;	5:6, 109:3,
0,2,-35,17:	[0,1,0,1] ;	11:4, 29:1, 109:3,
0,0,1,1:	[0,0,0,1] ;	7:1, 13:1, 19:2,
0,1,-20,70:	[0,1,0,0] ;	3:3, 5:6, 7:1, 13:1, 127:1,
0,1,-20,55:	[0,1,0,0] ;	3:6, 5:6, 19:1, 89:1,
0,0,2,65:	[0,0,0,0] ;	5:4, 7:1, 11:3, 17:1, 47:1, 59:1,
0,-1,8,74:	[1,0,0,0] ;	3:6, 11:6, 13:1, 67:1,
0,-2,5,5:	[1,1,0,1] ;	5:6, 11:4, 311:1,
0,-1,-5,30:	[1,1,0,1] ;	3:3, 5:6, 109:2, 113:1,
0,0,1,2:	[0,0,0,2] ;	3:3, 41:1, 127:1,
0,0,-2,37:	[1,0,0,0] ;	13:1, 17:2, 41:1, 43:1,
0,-1,-6,85:	[1,0,0,2] ;	3:3, 5:4, 89:1, 109:2, 373:1,
0,-1,-5,55:	[1,0,1,1] ;	5:6, 11:2, 29:1, 41:1, 331:1,
0,-1,-5,60:	[1,0,0,3] ;	3:3, 5:6, 67:1, 89:1, 541:1,
0,0,1,41:	[0,0,0,1] ;	3:3, 17:1, 29:1, 67:2, 229:1,
0,0,1,3:	[0,0,0,1] ;	43:1, 59:1, 167:1,
0,1,-20,10:	[1,1,0,0] ;	3:6, 5:6, 641:1,
0,1,-20,25:	[0,1,0,2] ;	3:3, 5:8, 193:1,
0,1,-19,25:	[0,1,1,3] ;	5:5, 11:4, 283:1
0,1,-19,75:	[0,0,1,1] ;	3:3, 5:5, 19:1, 41:1, 257:1,
0,-1,19,15:	[0,0,1,1] ;	3:3, 5:4, 17:1, 113:1, 347:1,
0,1,-18,75:	[0,0,0,2] ;	5:5, 29:2, 67:1, 191:1,
0,1,-17,55:	[0,1,1,3] ;	3:3, 5:4, 11:3, 307:1,
0,1,-8,10:	[0,1,0,2] ;	3:3, 5:4, 19:1, 43:1, 293:1,
0,-2,22,25:	[1,0,0,2] ;	3:6, 5:5, 17:1, 59:1, 353:1,
0,1,34,10:	[0,0,0,0] ;	3:6, 5:4, 41:1, 109:2, 349:1,
0,2,-45,22:	[1,1,0,0] ;	3:3, 11:2, 17:1, 79:1, 89:1, 109:2,
0,2,-45,100:	[1,1,0,0] ;	3:3, 5:8, 7:2, 29:1, 151:1,
0,2,-33,16:	[0,1,0,2] ;	3:3, 11:2, 71:1, 109:2, 227:1,
0,2,-25,12:	[0,1,0,0] ;	11:2, 13:1, 103:1, 109:2, 269:1,
0,2,-5,50:	[0,0,0,2] ;	5:8, 11:4, 19:1, 199:1,
0,-2,31,25:	[1,0,0,1] ;	3:6, 5:5, 43:1,7:1, 97:1
0,-2,35,20:	[1,0,0,2] ;	5:6, 17:1, 41:1, 89:1, 263:1,
0,-1,10,25:	[1,0,0,0] ;	5:8, 7:2, 41:1, 71:1,
0,-1,-33,25:	[1,1,1,3] ;	3:3, 5:5, 11:2, 19:1, 79:1, 89:1, 103:1,
0,-1,-16,90:	[1,1,0,0] ;	5:4, 7:1, 17:1, 43:1, 109:2, 197:1,
0,-1,-5,50:	[1,0,0,1] ;	5:8, 17:1, 97:1, 397:1,
0,-1,14,80:	[1,0,0,2] ;	3:3, 5:4, 13:1, 29:1, 89:1, 337:1,
0,0,-32,5:	[1,1,0,2] ;	3:3, 5:4, 11:2, 131:1, 197:1, 349:1,
0,0,-27,15:	[1,1,0,3] ;	3:6, 5:4, 7:2, 17:1, 103:1, 233:1,
0,0,-24,95:	[1,1,0,0] ;	5:4, 7:4, 89:1, 167:1, 271:1,
0,0,-16,85:	[1,0,0,0] ;	3:3, 5:4, 11:3, 89:1, 223:1,
0,0,-3,2:	[1,1,0,2] ;	223:1, 229:1, 367:1,
0,0,3,10:	[0,0,0,2] ;	5:4, 7:1, 307:1, 313:1,
0,1,-27,5:	[1,1,1,1] ;	3:3, 5:4, 11:3, 47:1, 479:1,
0,1,-20,35:	[0,1,0,2] ;	5:6, 7:2, 79:1, 181:1,
0,1,-8,82:	[0,0,0,0] ;	3:3, 7:1, 13:2, 17:1, 47:1, 349:1, 439:1,
0,1,-1,80:	[0,0,0,3] ;	5:4, 11:2, 17:1, 41:1, 541:1, 659:1,
0,2,-44,5:	[1,1,0,2] ;	5:4, 11:2, 313:2, 433:1,
0,2,-4,35:	[0,0,0,0] ;	3:3, 5:4, 67:1, 109:1, 151:1, 499:1,

0,-2,5,15:	[1,1,0,1] ;	3:3, 5:6, 509:1, 653:1,
0,0,-4,9:	[1,1,0,0] ;	13:1, 79:1, 449:1, 467:1,
0,1,-6,32:	[0,0,0,2] ;	3:3, 13:3, 59:1, 131:1, 499:1,
0,-2,-19,35:	[1,1,0,1] ;	5:4, 13:2, 17:2, 71:1, 131:1, 359:1,
0,-2,3,35:	[1,0,0,3] ;	3:3, 5:4, 17:2, 281:1, 331:1,
-2,19,-17,15:	[1,0,1,3] ;	3:3, 5:4, 11:2, 17:1, 23:3, 41:1, 479:1,
-2,16,-3,25:	[1,0,0,3] ;	3:3, 5:5, 31:3, 109:2, 233:1,
1,30,20,25:	[0,0,1,0] ;	5:10, 11:2, 37:3, 191:1,
-2,9,-15,35:	[1,1,1,1] ;	3:6, 5:6, 43:1, 53:3, 271:1,
-2,43,-95,-81:	[0,0,1,3] ;	3:6, 43:1, 61:3, 337:1, 659:1,
-2,44,-54,95:	[0,0,0,2] ;	5:4, 11:4, 13:1, 17:1, 73:3, 229:1,
-2,-47,-58,-70:	[1,1,0,0] ;	3:3, 5:4, 17:1, 19:1, 83:3, 89:1,229:1,307:1
1,-30,110,-75:	[1,1,1,0] ;	3:3, 5:10, 101:3, 109:1,
1,-47,-18,-30:	[1,1,0,1] ;	5:4, 17:1, 41:1, 107:3, 233:1, 577:1,
-2,11,-195,-125:	[1,1,1,3] ;	5:9, 11:3, 17:1, 127:1, 149:3,
-2,238,-235,50:	[0,0,0,2] ;	3:3, 5:8, 7:1, 11:3, 17:2, 157:3, 503:1,
-2,1,263,-175:	[1,0,1,1] ;	3:3, 5:5, 173:3, 307:1, 313:1, 563:1,
1,-4,-10,70:	[0,0,0,0] ;	3:3, 5:6, 47:1, 71:1, 179:3,
1,29,-75,-369:	[0,0,0,1] ;	3:3, 7:2, 17:2, 19:1, 103:1, 277:3, 433:1,
-8,85,16,90:	[1,0,0,2] ;	3:3, 5:4, 11:2, 67:2, 137:3, 421:1,
3,-61,101,-268:	[0,1,1,3] ;	3:6, 11:3, 41:1, 131:1, 139:3,
1,-54,341,45:	[1,0,1,2] ;	3:3, 5:4, 11:3, 13:1, 41:1, 131:1, 163:3,
-4,5,113,-495:	[1,1,1,1] ;	3:3, 5:4, 89:1, 199:1, 211:3, 293:1, 331:1,
3,52,-105,530:	[0,0,0,0] ;	3:3, 5:6, 11:3, 19:1, 191:1, 239:3, 263:1,
7,-290,330,-425:	[1,1,1,0] ;	3:3, 5:10, 17:1, 43:1, 131:1, 241:3, 353:1,
-6,106,65,60:	[1,0,0,0] ;	3:3, 5:6, 11:2, 223:1, 251:3,
33,-245,508,-20:	[0,0,0,3] ;	3:3, 54, 11:4, 3:2, 181:1,3173, 349:1,
0,-5,7,25:	[1,1,1,1] ;	3:6, 5:5, 13:1, 109:2, 197:1,
0,-2,15,15:	[1,0,0,1] ;	5:6,1:2, 47:2, 891:1
0,-1,-10,65:	[1,1,0,0] ;	3:6, 5:6, 41:1, 43:1, 67:1,
0,0,-6,87:	[1,0,0,2] ;	3:9, 47:1, 109:1, 199:1,
0,0,-1,22:	[0,0,0,0] ;	3:3, 17:1, 79:1, 89:1,

Table 4.4: The Values of ψ

For each rational prime p less than 360 the following table contains an entry

$$p : \psi(P)$$

where P is some prime in E lying over p of minimal degree. Thus P is of degree one unless p is a 3-prime. The value of ψ at P is independent of the choice of P unless p is a 2-prime. In this case the value of ψ given in the following table is determined by p only up to a sign (there are two primes of degree one over p, and according to the text the value of ψ on one is the negative of the value on the other). Note that the fourier coefficients of the L-series at 2-primes are 0 so that the values of ψ at 2-primes are not actually needed.

3 : -i	5 : 1	7 : -i	11 : 1	13 : 1	17 : -i
19 : i	23 : -i	29 : -1	31 : -i	37 : 1	41 : -i
43 : -i	47 : i	53 : 1	59 : i	61 : -1	67 : -1
71 : -i	73 : -1	79 : -i	83 : i	89 : -i	97 : -1
101 : 1	103 : -i	107 : -i	109 : -i	113 : -1	127 : 1
131 : i	137 : 1	139 : -i	149 : -1	151 : -i	157 : 1
163 : -i	167 : i	173 : 1	179 : i	181 : -i	191 : 1
193 : i	197 : -i	199 : i	211 : -i	223 : -1	227 : 1
229 : i	233 : -i	239 : -i	241 : 1	251 : -i	257 : -1
263 : i	269 : i	271 : -1	277 : 1	281 : -1	283 : i
293 : i	307 : -1	311 : -i	313 : i	317 : 1	331 : 1
337 : -1	347 : i	349 : i	353 : 1	359 : i	

Table 4.5: The L-series

Let $L(s,\mathbb{T}) = \Sigma a_n n^{-s}$ be the L-series of the icosahedral representation constructed in the text, so that a_n is 0 if n is not relatively prime to 2 and 5. For each integer n less than 360 and prime to 10 the following table contains an entry "n: a_n", where i denotes a fixed fourth root of unity and j is the positive root of $x^2-x-1 = 0$.

1: 1	3: -i	7: -ij	9: -1	11: 0	13: j	17: 0	19: i-ij
21: -j	23: -i	27: i	29: -1+j	31: -i	33: 0	37: -1	39: -ij
41: 0	43: -i+ij	47: ij	49: -1-j	51: 0	53: -1	57: 1-j	59: i-ij
61: 1	63: ij	67: 0	69: -1	71: -ij	73: 1	77: 0	79: -i+ij
81: 1	83: i	87: i-ij	89: 0	91: -i-ij	93: -1	97: -j	99: 0
101: -1	103:-i+ij	107: -i	109: 0	111: i	113: -1+j	117: -j	119: 0
121: 0	123: 0	127: 0	129: -1+j	131: ij	133: -1	137: -1	139: -i
141: j	143: 0	147: i+ij	149: 1	151:-i+ij	153: 0	157: -1	159: i
161: -j	163: -i	167: i-ij	169:1+i+j	171:-i+ij	173: -1	177: 1-j	179: i
181: 0	183: -i	187: 0	189: j	191: 0	193: 0	197: 0	199: ij
201: 0	203: -i	207: i	209: 0	211: -i	213: -j	217: -j	219: -i
221: 0	223: 0	227: 0	229: 0	231: 0	233: 0	237: -1+j	239: -i
241: -1	243: -i	247: 0	249: 1	251: -i	253: 0	257: -1+j	259: ij
261: 1-j	263: ij	267: 0	269: 0	271: 0	273: -1-j	277: -1	279: i
281: -1+j	283: ij	287: 0	289: 0	291: ij	293: 0	297: 0	299: -ij
301: 1	303: i	307: 0	309: -1+j	311: -ij	313: 0	317: -1	319: 0
321: -1	323: 0	327: 0	329: 1+j	331: 0	333: 1	337: -j	339: i-ij
341: 0	343: 2ij	347: i-ij	349: 0	351: ij	353: j	357: 0	359: ij

CHAPTER FIVE

If $\mathbb{T}:G_{\mathbf{Q}} \longrightarrow GL_2(\mathbf{C})$ is a galois representation then $\det(\mathbb{T}):G_{\mathbf{Q}} \longrightarrow \mathbf{C}^*$ is a one dimensional galois representation over $\mathbf{Q}$ and can therefore be identified with a Dirichlet character. The representation $\mathbb{T}$ is said to be odd if it takes complex conjugation to a matrix with eigenvalues 1 and -1, i.e.

$$\det\mathbb{T}(-1) = -1.$$

If $\mathbb{T}:G_{\mathbf{Q}} \longrightarrow GL_2(\mathbf{C})$ is an odd galois representation with Artin L-series

$$L(s,\mathbb{T}) = \Sigma\, a_n n^{-s} \quad (n \geq 1)$$

then define a function $f_{\mathbb{T}}$ on the upper half plane by

$$f_{\mathbb{T}}(q) = a_0 + \Sigma\, a_n q^n, \quad q = e^{2\pi iz}, \quad z \in \text{upper half plane}$$

$$a_0 = -L(0,\mathbb{T}).$$

The functional equation for $L(s,\mathbb{T})$ implies that $L(0,\mathbb{T})$ is nonzero if and only if $L(s,\mathbb{T})$ has a pole at $s = 1$. Thus $L(0,\mathbb{T})$ is nonzero if and only if $\mathbb{T}$ is a reducible representation of the form $1 \oplus \lambda$ (where 1 denotes the identity representation, which corresponds to the trivial Dirichlet character). The essential content of Theorem L-W of the introduction is that if $L(s,\mathbb{T}\otimes\lambda)$ is entire for all one dimensional representations λ then $f_{\mathbb{T}}$ is a modular form of weight one, level $\text{cond}(\mathbb{T})$, and character $\det(\mathbb{T})$.

The object of this chapter is to collect some results about modular forms that will be needed to demonstrate the existence of an icosahedral form of level 800. First we recall the known facts about those $\mathbb{T}$ for which the validity of

Artin's conjecture is "classical", namely imprimitive and reducible representations. Then we consider the case of dihedral forms (corresponding to irreducible properly induced representations) in detail and enumerate the possibilities for such forms whose level divides 800. This is then applied to the problem of finding a basis of the space of forms of weight two.

Reducible Representations

Theorem 7: Let χ_1 be an odd Dirichlet character of conductor N_1, and χ_2 be an even Dirichlet character of conductor N_2. Put $\mathbb{T} = \chi_1 \oplus \chi_2$. Then $f_{\mathbb{T}}$ is a modular form of weight one, level N_1N_2, and character $\chi_1\chi_2$. Its q-expansion is $\Sigma a_n q^n$ where

$$a_n = \sum_{m_1 m_2 = n} \chi_1(m_1)\,\chi_2(m_2) \qquad \text{if } n>0$$

$$a_0 = \begin{cases} 0 & \text{if } \chi_2 \text{ is nontrivial} \\ -L(0, \chi_1)/2 & \text{othwerise (i.e. if } N_2 = 1). \end{cases}$$

Remarks:

1) The modular forms produced by the above theorem lie in the space of Eisenstein series constructed by Hecke ([Hecke]) for the full congruence subgroup $\Gamma(N_1N_2)$. The proof of the theorem uses the usual relationships for reducible representations:

$$\det(\chi_1 \oplus \chi_2) = \chi_1\chi_2 \qquad \mathrm{cond}(\chi_1 \oplus \chi_2) = \mathrm{cond}(\chi_1)\mathrm{cond}(\chi_2)$$

$$L(s, \chi_1 \oplus \chi_2) = L(s, \chi_1)L(s, \chi_2)$$

2) For computational purposes the usual finite expressions for $L(0, \chi_1)$ can be used:

$$L(0, \chi_1) = (1/N_1)\,\Sigma\,\chi_1(n)n \quad = \quad 1/(\overline{\chi}_1(2)-2)\,\Sigma\,\chi(n)$$

where the first sum runs over all n prime to N_1 in the range $1 \leq n \leq N_1$, and the second sum runs over all n prime to N_1 in the interval $1 \leq n \leq N_1/2$.

The q-expansion of these Eisenstein series at cusps other than ∞ will be needed in the next chapter; these are described in appendix 6.

Induced Representations

In the next theorem the following notation will be used: $K/\mathbb{Q}$ is a quadratic extension with $G(K/\mathbb{Q}) = \{1,\sigma\}$ and $\Theta_{K/\mathbb{Q}}: C_{\mathbb{Q}} \longrightarrow \pm 1$ is the corresponding norm residue mapping. The character $\psi: C_K \longrightarrow \mathbb{C}^*$ is a nontrivial idele class character and $\mathbb{T} = \mathrm{Ind}(\psi, G_{\mathbb{Q}}/G_K)$ is the corresponding 2-dimensional galois representation.

Theorem 8: With the above notation, we have

$$\mathrm{cond}(\mathbb{T}) = D_{K/\mathbb{Q}} N_{K/\mathbb{Q}}(\mathrm{cond}(\psi))$$

and

$$\det(\mathbb{T}) = \Theta_{K/\mathbb{Q}} \quad \psi|C_{\mathbb{Q}}.$$

Hence $\mathbb{T}$ is odd if and only if either K is complex or K is real and ψ is trivial at exactly one of the real places of K. The representation $\mathbb{T}$ is reducible if and only if $\psi^{\sigma-1}$ is trivial, i.e. $\psi^{\sigma} = \psi$. If $\mathbb{T}$ is odd and irreducible then $f_{\mathbb{T}}$ is a cusp form. The modular form $f_{\mathbb{T}}$ determines K and the pair $\{\psi, \psi^{\sigma}\}$ unless $\psi^{\sigma-1}$ is of order 2, in which case $f_{\mathbb{T}}$ comes from 3 different quadratic fields K. If p is a rational prime that does not divide $\mathrm{cond}(\mathbb{T})$ then the p-th coefficient a_p of the q-expansion of $f_{\mathbb{T}}$ is

$$a_p = \begin{cases} 0 & \text{if p is inert in K} \\ \psi(P) + \psi(P^{\sigma}) & \text{if p splits as } PP^{\sigma} \text{ in K.} \end{cases}$$

The theorem follows from the usual facts about induced representations, including the relation

$$L(s,\mathbb{T}) = L(s,\psi) = \Pi(1-\psi(P)N_{K/\mathbb{Q}}(P)^{-s})^{-1}$$

where the product is over all prime ideals of K not dividing cond(ψ). Note that $L(s,\mathbb{T}) = L(s,\psi)$ is an abelian L-series and hence is holomorphic. The modular forms produced by the above theorem span the same space as Hecke's theta series of weight one associated to binary quadratic forms.

It is easy to enumerate the $f_{\mathbb{T}}$ associated to a pair of Dirichlet characters by theorem 7. Once some data representations are established it is also easy to have a computer churn out all of the possibilities for a particular level, for instance N = 800, and to then calculate the corresponding q-expansions.

The cusp forms $f_{\mathbb{T}}$ of Theorem 8 are another matter. It seems to be very difficult to enumerate the possibilities in general. For a specific level, say N= 800, the classification of the possible $f_{\mathbb{T}}$ is an exercise in the properties of some quadratic fields. We will briefly indicate the ideas used in this exercise, and then tabulate the results for levels dividing 800.

<u>Claim: In order to make cusp forms of level dividing 800 via Theorem 8 it suffices to look at the three quadratic fields</u>

$$K_1 = \mathbb{Q}(\sqrt{-1}) \quad K_2 = \mathbb{Q}(\sqrt{-2}) \quad K_3 = \mathbb{Q}(\sqrt{-5}).$$

Proof: In order for $f_{\mathbb{T}}$ to have conductor 800 the quadratic field K must be one of the 7 extensions of $\mathbb{Q}$ that is unramified outside 2 and 5. For such a K let $H \subset \Pi_P U_P \subset J_K$ be the maximal subgroup with the property that any character ψ of C_K with $\mathrm{cond}(\mathrm{Ind}(\psi))|800$ must vanish on H. Then the four fields other than the ones listed above can be eliminated for the following reasons:

$\mathbb{Q}(\sqrt{-10})$: $G(K/\mathbb{Q})$ acts trivially on J_K/K^*H so that there are no irreducible Ind(ψ) such that ψ vanishes on H.

$\mathbb{Q}(\sqrt{10})$: ditto; it is also easy to see that there are no ψ such that Ind(ψ) is odd.

$\mathbb{Q}(\sqrt{2})$: again there are no ψ with Ind(ψ) odd

$\mathbb{Q}(\sqrt{5})$: J_K/K^*H is a 2-group so that a representation π= Ind(ψ) coming from this field must also come from one of the three fields in the statement of the claim.

The proof of each of the above facts is a straightforward. q.e.d.

In order to describe the forms of weight one and conductor dividing 800 that come from the three quadratic fields above it is necessary to develop some notation to describe the idele class characters. The notation here is somewhat similar to the data representations chosen for the various programs using these modular forms; thus Table 5.1 was reproduced directly by the computer without any mistakes creeping in due to copying errors.

The Dirichlet characters that occur have conductor dividing 800. The abelian group of such characters is the product of three cyclic groups, so it is convenient to choose a basis for the group and then to describe any other character in terms of this basis. The group will be written additively.

Let a be the character of conductor 4 and order 2. Let b be the character of order 4 and conductor 16 such that

$$b(-1) = 1 \qquad b(3) = i = \exp(\pi i/2).$$

Let c be the character of order 20 and conductor 25 such that

$$c(2) = \exp(2\pi i/20) \text{ (2 is a primitive root mod 25)}.$$

From now on we will denote the character $a^i b^j c^k$ by [i,j,k]. Any Dirichlet character of conductor dividing 800 whose values lie in the field of 20-th roots of unity is of this form.

In the field $K_1 = \mathbf{Q}(\sqrt{-1})$, 2 is ramified and 5 splits. If ψ is an idele class character on K_1 such that $\mathrm{cond}(\mathrm{Ind}(\psi))|800$ then ψ has conductor with exponent at most 3 at the prime above 2 and at most 2 at each of the primes above 5. The condition that $\psi(i) = 1$ (here i is the principal idele) can be shown to imply that the component of such a ψ at 2 is uniquely determined by the components of ψ at the two primes above 5. Since the class number of K_1 is 1 this means that a character ψ is completely specified by giving its components on the units at the two primes above 5. Let $(5) = PP'$ and let 2_P (resp. $2_{P'}$) be the idele that is 2 at P (resp. P') and 1 outside P (resp. P'). Let a be the character of conductor P^2 such that $a(2_P) = \exp(2\pi i/20)$ (recall that 2 is a primitive root mod 25) and let b be the character of conductor P'^2 with $b(2_{P'}) = \exp(2\pi i/20)$. From now on [i,j;1] will denote the idele class character $a^i b^j$ of K_1. The nontrivial element σ of $G(K_1/\mathbf{Q})$ interchanges the primes above 5. The action of $G(K_1/\mathbf{Q})$ on the ideles of K_1 induces an action of $G(K_1/\mathbf{Q})$ on the idele class characters and σ takes [i,j;1] to [j,i;1].

In the field $K_2 = \mathbf{Q}(\sqrt{-2})$ 2 is ramified and 5 is inert. Again the class number is one so that it suffices to specify a character on the units. If ψ is an idele class character on K_2 such that $\mathrm{cond}(\mathrm{Ind}(\psi))|800$ then ψ must have conductor of exponent at most 2 at the prime above 2 and at most 1 at the prime above 5. The number $1+\sqrt{-2}$ generates the residue field at the prime P above 5; let x_P be the idele that is $1+\sqrt{-2}$ at P and 1 elsewhere. Let a be the unique character of the ideles whose conductor is the square of the prime above 2 and let b be the character of conductor P such that $b(x_P) = \exp(2\pi i/24)$. From now on [i,j;2] will denote the character $a^i b^j$. In order for [i,j;2] to be an idele class character it suffices that it vanish on -1 which means that j must be even. The nontrivial element of $G(K_2/\mathbf{Q})$ acts trivially on the first component and as multiplication by 5 on the second component.

In the field $K_3 = \mathbf{Q}(\sqrt{-5})$ 2 and 5 both ramify. Unfortunately the class number is 2. Any character ψ with $\mathrm{cond}(\mathrm{Ind}(\psi))|800$ must have conductor of exponent at most 3 at the prime above 2 and at most 1 at the prime above 5. Let x be the idele which is $1+\sqrt{-5}$ at the prime above 2 and 1 elsewhere. Let y be the idele that is 3 at the prime above 5 and 1 elsewhere. It can be shown that x and y generate $J_{K_3}/K_3{*}H$. Let a be the character that takes x to i and takes y to 1. Let b be the character that takes x to 1 and takes y to i. From now on let [i,j;3] denote the character $a^i b^j$.

In terms of the concrete notation above it is now necessary to 1) determine which characters are in fact idele class characters, 2) find the action of galois on the characters so that the ψ fixed by galois (which give reducible Ind(ψ)) can be eliminated, 3) eliminate duplications (e.g. Ind(ψ)= Ind(ψ^σ); also some representations come from both K_3 and K_1). This amounts to a straightforward if not particularly illuminating exercise; the results are given in Table 5.1.

Forms of Weight 2

The above information yields modular forms of type $(1,\varepsilon,800)$ for various odd Dirichlet characters ε. For the calculations described in the next chapter it is necessary to have a basis for the vector space $S^2(800)$ of forms of weight 2 on $\Gamma_0(800)$, i.e., forms of type (2,1,800).

It is natural to try to obtain such forms by multiplying together forms of type $(1,\varepsilon,800)$ with forms of type $(1,\varepsilon^{-1},800)$. This procedure was tried for all levels dividing 800. The pool of forms of weight one was taken to be the cusp forms tabulated above together with those Eisenstein series whose q-expansion coefficients were in the field of 20-th roots of unity. To insure that only cusp forms were obtained one of the multiplicands was always a cusp form (though from the results given in appendix 6 on the values of Eisenstein series at the cusps it is clear that this is unnecessarily conservative - often the product of two Eisenstein series is a cusp form). Thus it is obvious that this process will not produce a basis of the forms of weight 2 for the small levels since the cusp form of weight one with lowest conductor has conductor 80. However this procedure was successful for levels 100,200,400 and 800. The dimensions of the various spaces involved are contained in Table 5.2. As a byproduct of this procedure one also obtains the "gap sequence" of the different Riemann surfaces at the cusp ∞; this information is also in Table 5.2.

Remarks: On the basis of this limited experimental evidence it seems reasonable to suppose that the above procedure is promising when the level is highly composite (so that there are many quadratic fields and idele class characters). The actual implementation of these ideas involves several interesting problems, both in the choice of data representations and in the choice of algorithms. The computation time for the entire process above is negligible.

All of the calculations were done with 400 terms of the q-expansions of the various modular forms. The fact that the computer finds a 97-dimensional space of forms of weight 2 and level 800 is in itself very strong evidence that the machine is calculating correctly. From a naive numerical point of view it is a miracle that all of the (approximately 1000) products of forms of weight one all lie in a 97-dimensional subspace of the underlying 400 dimensional space. Moreover, it was easy to check that this 97-dimensional space was stable under the Hecke operators U_2 and T_3.

Table 5.1

The following is a list of ψ that correspond to dihedral forms of weight one and level dividing 800; the notation is described in the text.

ψ	ψ^{σ}	det(Ind(ψ))	cond(Ind(ψ))
[0, 1; 1]	[1, 0; 1]	[0, 0, 1]	800
[0, 2; 1]	[2, 0; 1]	[1, 0, 2]	400
[0, 3; 1]	[3, 0; 1]	[0, 0, 3]	800
[0, 4; 1]	[4, 0; 1]	[1, 0, 4]	100
[0, 5; 1]	[5, 0; 1]	[0, 0, 5]	160
[0, 6; 1]	[6, 0; 1]	[1, 0, 6]	400
[0, 7; 1]	[7, 0; 1]	[0, 0, 7]	800
[0, 8; 1]	[8, 0; 1]	[1, 0, 8]	100
[0, 9; 1]	[9, 0; 1]	[0, 0, 9]	800
[5,10; 1]	[10, 5; 1]	[0, 0,15]	800
[0,10; 1]	[10, 0; 1]	[1, 0,10]	80
[0,11; 1]	[11, 0; 1]	[0, 0,11]	800
[0,12; 1]	[12, 0; 1]	[1, 0,12]	100
[0,13; 1]	[13, 0; 1]	[0, 0,13]	800
[0,14; 1]	[14, 0; 1]	[1, 0,14]	400
[10,15; 1]	[15,10; 1]	[0, 0, 5]	800
[5,15; 1]	[15, 5; 1]	[1, 0, 0]	400
[0,15; 1]	[15, 0; 1]	[0, 0,15]	160
[0,16; 1]	[16, 0; 1]	[1, 0,16]	100
[0,17; 1]	[17, 0; 1]	[0, 0,17]	800
[0,18; 1]	[18, 0; 1]	[1, 0,18]	400
[0,19; 1]	[19, 0; 1]	[0, 0,19]	800
[0, 2; 2]	[0,10; 2]	[1, 2,10]	200
[0, 4; 2]	[0,20; 2]	[1, 2, 0]	200
[0, 8; 2]	[0,16; 2]	[1, 2, 0]	200
[0,14; 2]	[0,22; 2]	[1, 2,10]	200
[1, 2; 2]	[1,10; 2]	[1, 2,10]	800
[1, 4; 2]	[1,20; 2]	[1, 2, 0]	800
[1, 8; 2]	[1,16; 2]	[1, 2, 0]	800
[1,14; 2]	[1,22; 2]	[1, 2,10]	800
[0, 1; 3]	[3, 1; 3]	[0, 0, 5]	800
[1, 1; 3]	[2, 1; 3]	[0, 0, 5]	800
[0, 3; 3]	[1, 3; 3]	[0, 0,15]	800
[2, 3; 3]	[3, 3; 3]	[0, 0,15]	800

Table 5.2

Modular Forms of Weight 2 and Level 800

The following tables contain some information on the spaces of modular forms of weight two and level dividing 800. The dimensions for the untabulated levels that divide 800 are all 0. Definition: a positive integer m is a gap for a Riemann surface at a point x if there is a differential with a zero of order m-1 at that point. In the tables below the Riemann surfaces are the $X_0(N)$ for N = 100,200,400, and 800, and the point is ∞. A cusp form of weight two corresponds to a differential of the first kind and if the first non-zero term of the q-expansion of the form is the m-th then the differential has a zero of order m-1. The asterisks correspond to gaps in the gap sequence.

N:	dim $S_2(N)$	dim $S_2^{new}(N)$
20	1	1
32	1	1
40	3	1
50	2	2
80	7	2
100	7	1
160	17	4
200	19	5
400	43	5
800	97	19

Level:	gap sequence
100	1, 2, 3, 4, 5, *, 7, *, 9
200	1, 2, 3, 4, 5, 6, 7, 8, 9,10,11, *,13,14,15, *, 17,18,19, *,21, *,23
400	1, 2, 3, 4, 5, 6, 7, 8, 9,10,11,12,13,14,15,16, 17,18,19,20,21,22,23, *,25,26,27,28,29,30,31, *, 33,34,35,36,37,38,39, *, *,42,43, *, *,46, *, *, 49, *,51, *,53
800	1, 2, 3, 4, 5, 6, 7, 8, 9,10,11,12,13,14,15,16, 17,18,19,20,21,22,23,24,25,26,27,28,29,30,31,32, 33,34,35,36,37,38,39,40,41,42,43,44,45,46,47, *, 49,50,51,52,53,54,55,56,57,58,59,60,61,62,63, *, 65,66,67,68,69,70,71,72,73,74,75,76,77,78,79, *, 81, *,83,84,85,86,87, *,89, *, *,92,93, *, *, *, 97,98,99, *,101,102,103,*,105,106,*,*,109,*,*,*, *,*, ... *,129,*,*,*,133

CHAPTER SIX

Using the notation of the previous chapter define two Dirichlet characters μ, ν by

$$\mu = [1,0,0], \nu = [0,0,4];$$

μ is the unique character of order 2 and conductor 4 and ν is a character of order 5 and conductor 25.

Let V be the vector space of modular forms of type $(1, \mu\nu, 800)$. The Hecke operators T_p $(p \neq 2,5)$ act on V. If a cusp form f in V is an eigenform for all of these Hecke operators then Theorem D-S of the introduction implies that there is an irreducible odd two-dimensional galois represention $\mathbb{T}$ over $\mathbb{Q}$ with conductor dividing 800 and determinant equal to $\mu\nu$ such that

$$L_f(s) = L(\mathbb{T}, s).$$

We say that an eigenform f is dihedral if the corresponding representation $\mathbb{T}$ is an irreducible properly induced representation and that f is tetrahedral (resp. octahedral, resp. icosahedral) if $\mathbb{T}$ is tetrahedral (resp. octahedral, resp. icosahedral).

There are two non-cuspidal eigenforms of level 100 in V; the corresponding reducible representations are

$$1 \oplus \mu\nu \quad \text{and} \quad \mu \oplus \nu \quad .$$

Let g_1 and g_2 be the corresponding modular forms. Each of these forms can be "pushed up" ([Atkin-Lehner]) by the operators B_1, B_2, B_4, and B_8 to give a form

$$g_i | B_d \in V.$$

It is easy to check that there are no other pairs of Dirichlet characters whose product is $\mu\nu$ and the product of whose conductors divides 800. Hence V is the direct sum of the space of cusp forms in V with the space spanned by the $g_i|B_d$, $i = 1,2$, $d = 1,2,4,8$.

There is only one dihedral form of character $\mu\nu$ in Table 5.2. The representation comes from the idele class character on $\mathbf{Q}(\sqrt{-1})$ that "is" ν on the units at one of the primes above 5 and is unramified elsewhere. Again the representation has conductor 100. Hence the modular form, which we will denote by g_3, generates a four dimensional subspace of V via the operators B_1, B_2, B_4, and B_8.

There are no cyclic extensions of $\mathbf{Q}$ of degree 3 unramified outside 2 and 5; hence there are no A_4 extensions of $\mathbf{Q}$ unramified outside 2 and 5. It follows that there are no tetrahedral representations of conductor dividing 800 and hence (Theorem D-S) no tetrahedral forms in V.

According to the results in appendix 4 there are only three S_4 extensions of $\mathbf{Q}$ unramified outside 2 and 5 and the corresponding representations have conductors that do not divide 800. So there are no octahedral forms in V.

<u>Theorem 9: There is an icosahedral form in</u> V.

<u>Corollary: There is an icosahedral representation</u> $\pi_0 : G_{\mathbf{Q}} \longrightarrow GL_2(\mathbf{C})$ <u>of conductor</u> 800 <u>and determinant</u> $\mu\nu$ <u>such that all twists of</u> π_0 <u>have holomorphic L-series</u>.

The Corollary follows immediately from the theorem via Theorem D-S.

We start the proof of the theorem by considering the icosahedral representation constructed in chapter 4. To fix the representation make the following specific choice of u and u':

$$u = (3+\sqrt{5})/2 \qquad u' = (3-\sqrt{5})/2.$$

This means that the frobenius of a prime of type 5A (resp. 5B) in Table 4.1 has

$tr(F_p)^2/det(F_p)$ equal to u (resp. u'). Call this representation $\mathbb{T}$, so that $det(\mathbb{T}) = \mu$. Let $\mathbb{T}_1 = \mathbb{T} \otimes \nu^3$. Let L(s) be the Dirichlet series obtained by multiplying together the p-Euler factors of $L(s,\mathbb{T}_1)$ for $p \neq 5$ and then multiplying by the Euler factor

$$(1 - \nu(2)^2 5^{-s})$$

at 5. Let f be the power series in q obtained by taking the "Mellin transform" of L(s) (as in the definition of $f_{\mathbb{T}}$ in the previous chapter).

In order to express the next assertion concisely we introduce some notation. If g_i is one of the three forms introduced above then let $g_{i,d} = g_i | B_d$, for i= 1,2,3 and d= 1,2,4,8. If g is a modular form of type $(1,\epsilon,N)$ then let $\bar{g}$ denote the "complex conjugate" of g so that $\bar{g}$ is a form of type $(1,\bar{\epsilon},N)$ and the coefficients of the q-expansion of $\bar{g}$ are the complex conjugates of the coefficients of the q-expansion of g If the first M terms of two power series in q agree then we say that they are congruent modulo q^M.

Fact: For each i= 1,2,3 and d= 1,2,4,8 there is a modular form $h_{i,d}$ of type (2,1,800) such that

$$f\bar{g}_{i,d} \equiv h_{i,d} \bmod q^{360}.$$

Remarks: 1) If $\mathbb{T}|G_5 \cong \nu^2 \oplus \nu^{-2}$ (where G_5 is a deomposition group at 5) then $cond(\mathbb{T}_1) = 800$ and one can then use the definition of the Artin L-functions at ramified primes to show that the 5-Euler factor above is the 5-Euler factor of $L(s,\mathbb{T}_1)$, so that f is the Mellin transform of $L(s,\mathbb{T}_1)$. If this is not true (i.e. $cond(\mathbb{T}_1) = 20000$ which is to say that $\mathbb{T} | G_5 \cong \nu^{-1} \oplus \nu$) then then we could interchange u and u' above to make $cond(\mathbb{T}_1) = 800$; we make the specific choice above because the above Fact is then true. However, the proof of Theorem 9 is independent of any assumptions about the conductor of $\mathbb{T}_1$; we start with a power series f (which could in principle be obtained by black magic) and proceed to show that f coincides with the q-expansion of a modular form of weight one to 360 places.

2) The information and tools necessary to routinely verify this basically numerical assertion have been presented in the earlier chapters; unfortunately the amount of computation required may be somewhat more than even the most energetic reader could stomach. The techniques that I used to verify this using a computer are briefly touched upon in Appendix 5. It is virtually impossible to present any evidence for the validity of the above Fact that would be readily susceptible to human verification. For instance, one could write the left hand side above as a linear combination of the basis elements of the space of forms of weight two obtained in chapter 5 and then tabulate these coefficients; however the resulting table would be virtually meaningless to the human eye. Just as a proof should not be accepted until it has been confirmed independently by different mathematicians the ultimate test of the above claim lies in independent verification; here this has something of the flavor of an experiment in the physical sciences in that here this verification requires the use of a complicated machine, namely a computer. However, the mere fact that the computer purports to check this claim is very strong evidence for its validity! From a strictly numerical point of view the left hand side of the congruence is just the first 360 terms of some power series, and to find that this power series lies in a 97-dimensional subspace is incredibly convincing evidence that something highly non-random is happening.

According to the above Fact,

$$h_{i,d}\overline{g}_{j,d'} \equiv h_{j,d}\overline{g}_{i,d} \bmod q^{360}, \quad \text{for all } i,j,d,d'.$$

On each side of this congruence we have modular forms of type $(3, \mu\,\overline{\nu}, 800)$. Modular forms of this type are sections of a bundle on $X_0(800)$ whose degree is 344 (see Appendix 6 or [Tate, NFZ]; 360 is a trivial upper bound for the degree). Thus the modular forms in the above congruence must actually be equal since their difference has a zero at ∞ of order bigger than 360. Therefore we can unambiguously define a function

$$f' = h_{i,d}/\bar{g}_{i,d}$$

that is independent of the choice of i and d. If the function f' is holomorphic then it is a modular form of weight one and character $\mu\nu$; i.e. f' satisfies all the requirements for such a modular form except that possibly it has poles at points where all of the $\bar{g}_{i,d}$ vanish. We will show that in fact f' is holomorphic by showing first that the forms $\bar{g}_{i,d}$ have no zero in the upper half plane and then showing that f' does not have a pole at any of the cusps.

Proposition: The modular forms $\bar{g}_1$, $\bar{g}_2$, and $\bar{g}_3$ have no common zero in the upper half plane.

Proof: The first 10 terms of the q-expansions of the $\bar{g}_i$ are as follows:

n :	0	1	2	3	4	5	6	7	8	9	10
$\bar{g}_1$:	z^4-z	1	1	$1-z^2$	1	1	$1-z^2$	0	1	z^4-z^2+1	1
$\bar{g}_2$:	0	1	z	z^2-1	z^2	1	z^3-z	0	z^3	z^4-z^2+1	z
$\bar{g}_3$:	0	1	z^3	0	z	z^4	0	0	z^4	z^2	z^2

where $z = e^{2\pi i/5}$. Using this table it is easy to verify that the following 5 forms

$$\bar{g}_1 g_3 \quad \bar{g}_2 g_3 \quad \bar{g}_3 g_1 \quad \bar{g}_3 g_2 \quad \bar{g}_3 g_3$$

are linearly independent so that they generate a 5-dimensional subspace of the 7 dimensional vector space $S^2(100)$ of cusp forms of type (2,1,100) (which can be identified with the space of differentials of the first kind on the Riemann surface $X_0(100)$). This is the first of several numerical assertions in this proof that will be left to the reader; it is possible, if boring, to verify each of these statements by hand computations.

Note that the 5 forms above are automatically cusp forms since g_3 and $\bar{g}_3$ are cusp forms Using the results of Appendix 6 one can show that $g_1\bar{g}_2$ is a cusp form; however $g_1\bar{g}_2$ lies in the space spanned by the above forms so this fact is not useful here.

According to [Antwerp IV, p. 137] $S^2(100)$ decomposes as follows under the "partial" W-operators W_2 and W_5:

W_2 eigenvalue:	+1	+1	-1	-1
W_5 eigenvalue:	+1	-1	+1	-1
dimension:	1	1	3	2.

It is possible to check that the space generated by the five forms above contains the 2 dimensional space of forms fixed by W_2. This can be done by a direct construction of the space of forms of weight 2 for which one knows the action of W_2 (à la Tingley, see [Miller]) or by using a formula for the action of partial W-operators due to Dan Flath, which here says that

$$g_1|W_2 = -i\ g_2;\ g_2|W_2 = -\ g_1;\quad g_3|W_2 = -i\ g_3.$$

The differentials on $X_0(100)/W_2$ can be identified with the modular forms of type (2,1,100) that are fixed by W_2. Differentials on a Riemann surface have no common zero, and so if $\bar{g}_1$, $\bar{g}_2$, and $\bar{g}_3$ have a common zero in the upper half plane then the common zero must be a fixed point of W_2 (there are no elliptic points so the upper half plane is unramified over $X_0(100)$).

Using the ideas in [Ogg, HC, p. 453] it is easy to see that there are 2 fixed points of W_2 in the upper half plane, and that they can be taken to be

$$(\pm 36 + 2i)/1300.$$

Call these points z_1 and z_2. They correspond to the elliptic curve with complex multiplication by the order generated by 1 and 2i in $\mathbf{Z}[\sqrt{-1}]$). With this explicit realization of the fixed points it is possible to check that

$$\bar{g}_i(z_j) > 1 \quad \text{for all } i = 1,2,3 \text{ and } j = 1,2$$

so that the $\bar{g}_i$ certainly do not have a common zero at the z_j. q.e.d

An alternative proof to the above proposition could be made following [Tate, N133] more closely by showing (for instance) that if h' and h'' span the part of the space of differentials on $X_0(100)$ not spanned by multiples of the $\bar{g}_i$, then h'^2 and h''^2 are multiples of the $\bar{g}_i$ (or that some power of h' and h'' lies in the ideal generated by the $\bar{g}_i$ in the graded ring of modular forms for $_0(100)$).

Now we must show that f' has no poles at the cusps. For this we use the notation and results of Appendix 6, which is concerned with q-expansions of modular forms at arbitrary cusps. Unless explicitly stated otherwise, any reference to a lemma, theorem or corollary in the remainder of the proof of Theorem 9 is a reference to the corresponding result in appendix 6.

Recall that if p = [x,y] is a cusp of $X_1(N)$ then d = (y,N) is called the divisor of the cusp. To show that f' has no poles at the cusps we will explicitly examine each cusp of $X_0(800)$; the technique required to show that f' is holomorphic at a given cusp basically depends only on the divisor of the cusp. The number of cusps with a divisor d is $\varphi(d,N/d)$; in all there are 48 cusps on $X_0(800)$.

According to corollary 1, $\bar{g}_1$ (resp. $\bar{g}_2$) is nonzero at precisely two cusps of $X_0(100)$: [1,0] and [0,1] (resp. [1,4] and [1,25]). Thus $\bar{g}_{1,1}$ (resp. $\bar{g}_{2,1}$) does not vanish at any cusp of $X_0(800)$ lying over [1,0] or [0,1] (resp. [1,4] and [1,25]) and hence $f' = h_{1,1}/\bar{g}_{1,1} = h_{2,1}/\bar{g}_{2,1}$ does not have a pole at any of these cusps.

Any cusp of $X_0(800)$ of divisor 800, 400, 200 or 100 lies over the cusp [1,0] of $X_0(100)$. The unique cusp of $X_0(800)$ of divisor 1 (resp. 25) lies over the cusp

[0,1] (resp. [1,25]) of $X_0(100)$. The cusps of $X_0(800)$ of divisor 4,8,16 or 32 lie over the cusp [1,4] of $X_0(100)$. Using Lemma 4 we see that $\overline{g}_{1,2}$ (resp. $\overline{g}_{2,2}$) does not vanish at [1,50] (resp. [1,2]).

This accounts for all of the cusps of $X_0(800)$ except for the cusps of divisor 5,10,20,40,80 or 160. By using corollary 2 we can compute that

1) $v_p(g_1) < 1$ if p is any cusp of $X_0(100)$ of divisor 20

2) $v_p(g_1) < 1/4$ if p is any cusp of $X_0(100)$ of divisor 5

(the notation is as in appendix 6; $v_p(g)$ is the order of zero of g at the cusp p measured in terms of the translation of the usual uniformizing parameter q at ∞).

Any cusp of $X_0(800)$ of divisor 40, 80, or 160 has ramification index 1 over and moreover lies over a cusp of $X_0(100)$ of divisor 20. Therefore for such a cusp

$$v_p(\overline{g}_1) < 1 \leq v_p(h_{1,1})$$

so $f' = h_{1,1}/\overline{g}_1$ does not have a pole at such a cusp p.

This leaves the cusps of $X_0(800)$ of divisors 5, 10 and 20. For each such cusp lemma 1 together with 2) above show that

$$v_p(g_{1,8}) < 1/e \leq v_p(h_{1,8})$$

($e = N/d(d,N/d)$ is the ramification of p over ∞) so f' does not have a pole at these cusps. More precisely:

If p is a cusp of divisor 5 then in Lemma 4 we take $f = \overline{g}_1$, and $d = 8$. Then p' is a cusp of $X_0(100)$ of divisor 5 and $r = 1$ so that

$$v_p(\overline{g}_{1,8}) = (1/8)v_{p'}(\overline{g}_1) < 1/32 = 1/e$$

(recall that 2) above says that $v_p(\overline{g}_1)<1/4$).

If p is a cusp of divisor 10 then in Lemma 4 we take $f = \overline{g}_1$ and $d = 8$. Then p' is a cusp of $X_0(100)$ of divisor 5 and $r = 2$ so that

$$v_p(\overline{g}_{1,8}) = (4/8)v_{p'}(\overline{g}_1) < 1/8 = 1/e.$$

If p is a cusp of divisor 20 then we have $d = 8$, $r = 4$ and hence

$$v_p(\overline{g}_{1,8}) = (16/8)v_p(\overline{g}_1) < 1/2 = 1/e.$$

This finishes the proof that f' is a holomorphic modular form.

Remarks: The proof that the "multipliers" $\overline{g}_i$ have no common zero in the upper half plane was even easier than the proof in [Tate, N133] for the case of the tetrahedral representation of conductor 133. However the difficulties at the cusps are not present for this tetrahedral representation since for square free level there is always an Eisenstein series that does not vanish at some specified cusp. Thus the highly composite nature of the level 800 (which seems to be a necessary condition for finding a low icosahedral conductor and which seems to make it easy to construct a basis of the space of forms of weight 2) causes difficulties at the cusps.

The last step of the proof of Theorem 9 is to show that the fact that f' is a modular form implies that V must contain an icosahedral form.

The power series f was defined by an Euler product and so f is (formally) an eigenform of the Hecke operator T_3 (with eigenvalue $\beta = e^{6\pi i/20}$). The modular form f' is congruent to f modulo q^{360} so

$$T_3 f' \equiv T_3 f \equiv \beta f \equiv \beta f' \bmod q^{120}.$$

The degree of the bundle of cusp forms of type $(1,\mu\nu,800)$ is 86 (again this is a consequence of the results stated in appendix 6; 120 is a trivial upper bound). Therefore f' must be an eigenform for the Hecke operator T_3, and have eigenvalue β. This eigenvalue does not occur as a T_3-eigenvalue for any of the non-cuspidal or dihedral forms in V. There are no tetrahedral or octahedral forms in V. Hence there must be an icosahedral form in V!!! q.e.d.

This proof is somewhat disingenuous, since it evades interesting questions at each stage (e.g., is f' an eigenform for all of the Hecke operators? Is the representation $\mathbb{T}_0$ of the corollary the same as the original icosahedral representation $\mathbf{T}_1$? Is f the Mellin transform of the L-series of $\mathbf{T}_1$? .....)

The next theorem says that if f' is an eigenform for the Hecke operator T_{11} then the above questions have affirmative answers. The proof uses the techniques of Odlyzko, Serre, and Poitou for bounding discriminants (we will use [Poitou] as the reference for this material); the idea is essentially to show that the galois representation of conductor 800 is unique.

In order to clarify the statement of the theorem it is convenient to review some of the notation accumulated during the proof of Theorem 9:

$\mathbb{T}$ is the icosahedral representation constructed in chapter 4,

$\det(\mathbb{T}) = \mu$

$\mathbf{T}_1 = \mathbb{T}\otimes\nu^3$; the determinant of $\mathbf{T}_1$ is $\mu\nu$

f is the power series derived from $L(s,\mathbb{T}_1)$ except that we force f to have the 5-Euler factor that it would have if $\mathrm{cond}(\mathbb{T}_1) = 800$ (which is equivalent to saying that $\mathbb{T}|G_5 \equiv \nu^2\oplus\nu^3$)

f' is a modular form of weight one and character $\mu\nu$; we know that f' is an eigenform for T_3 and that its q-expansion agrees with f to 360 terms

$\mathbf{T}_0$ is an icosahedral representation that comes from an icosahedral form in V so that $\mathbf{T}_0$ satisfies Artin's conjecture;

we do not know that f' is the corresponding modular form since we haven't shown that f' is a Hecke eigenform.

Theorem 10: Assume that f' is an eigenform of T_{11}. Then

1) f' is an eigenform for all T_p, $p \neq 2,5$

2) $\mathbf{T}_1$ and all of its twists satisfy Artin's conjecture

3) $\mathbb{T}_0 = \mathbb{T}_1$

4) cond($\mathbf{T}_1$) = 800, so that f is the Mellin transform of $L(s,\mathbf{T}_1)$.

The theorem follows in a straightforward fashion from the following statement:

Proposition: There is a unique icosahedral galois representation of conductor 800, determinant $\mu\nu$, and

$$\mathrm{Trace}(F_3) = e^{6\pi i/20}, \quad \mathrm{Trace}(F_{11}) = 0$$

where $F_p \in GL_2(\mathbf{C})$ is the image of a frobenius element at p.

Remark: By the calculations in chapter 4 the last two conditions in the Proposition imply that 3 (resp. 11) is a 3-prime (resp. 2-prime) in the corresponding A_5 field.

First we outline the proof of the theorem using the proposition and then we prove the proposition.

It is straightforward to check that f' is of level 800 (and not of some lower level) by, for instance, checking that the "Fact" above is not true if f is replaced by any push-up of f. Since we know that the modular form f' is an eigenform of T_3 (whose eigenvalue is such that 3 is a 3-prime in the corresponding icosahedral field) and we are assuming that f' is an eigenform of T_{11} (with eigenvalue = $a_{11} = 0$) it follows that the representation $\mathbf{T}_0$ satisfies the conditions of the proposition.

Let σ be an automorphism of **C** that interchanges u and u'. Then either $\mathrm{cond}(\mathbf{T}_1) = 800$ or $\mathrm{cond}(\mathbf{T}^{\sigma} \otimes \nu^3) = 800$. The "Fact" above is false if $\mathbf{T}_1 = \mathbb{T} \otimes \nu^3$ is replaced by $\mathbf{T}_1 = \mathbb{T}^{\sigma} \otimes \nu^3$; thus the latter representation would not satisfy Artin's conjecture. If the latter representation has conductor 800 then it satisfies the conditions of the Proposition; by the uniqueness assertion of the Proposition and the fact that $\mathbf{T}_0$ satisfies Artin's conjecture we are forced to conclude that $\mathrm{cond}(\mathbf{T}_1) = 800$ and that $\mathbf{T}_1 = \mathbf{T}_0$. This proves 2), 3) and 4) of the theorem.

These assertions of the theorem show that f' must be the Mellin transform of $L(s,\mathbf{T}_1)$, and that $\mathbf{T}_1$ and all of its twists must satisfy Artin's conjecture. By Theorem L-W, f' must be an eigenform of all of the Hecke operators. This finishes the proof of the theorem.

Proof of the proposition: The representation $\mathbf{T}_1$ satisfies the conditions of the proposition.

We start the proof of uniqueness by stating a bound for the discriminant of a number field; this inequality is essentially lifted straight out of [Poitou].

If d is the discriminant of a totally complex algebraic number field of degree n over **Q**, then one can apply "explicit formulae" to the zeta function of the field and the auxiliary function

$$F(x) = e^{-yx^2}/\cosh(x/2)$$

(where y is a parameter that will be chosen below) to obtain

$$*)\qquad (1/n)\log d \geq \gamma + \log 4 - 8.6n^{-2/3} + (2/n)\Sigma\, C(m,P).$$

The sum runs over all positive integers m and all prime ideals P of the field, and

$$C(m,P) = F(\log NP^m)\ \log(NP)/NP^{m/2}$$

where N denotes thenorm from the field down to $\mathbf{Q}$ and the auxiliary parameter y is chosen to be

$$y = 1/4b, \quad b^{3/2} = 1.0518n/2\pi \quad .$$

For a proof and a more detailed discussion see [Poitou]; the inequality *) above is essentially formula (22) in [Poitou] except that 1) we are dealing with a totally complex field, 2) we have retained the term that corresponds to the logarithmic derivative of the zeta function (in formula (12) in [Poitou]) and 3) we have made the choice of the parameter b explicit.

Suppose that there are two distinct representations satisfying the conditions in the proposition. Let K and K' be the corresponding A_5 extensions of $\mathbf{Q}$. If K = K' then by the theory in chapter 4 and the constraint on the conductor and determinant, the two representations must differ by a character of order 2 that is unramified outside 2 and 5. Since we also specify the eigenvalue at 3 it follows that if K = K' then the two representations must be isomorphic. Thus the proposition will be proved if we can show that $K \neq K'$ leads to a contradiction; we will do that by applying the inequality *).

From now on assume that $K \neq K'$ and let L be the compositum KK'. Since A_5 is a simple group and the intersection of K and K' is a galois extension of $\mathbf{Q}$ we see that this intersection must be $\mathbf{Q}$ which means that the galois group of L over $\mathbf{Q}$ must be $A_5 \times A_5$.

It is easy to use Table 3.1 to check that the A_4 extension of $\mathbf{Q}_2$ is the unique extension of $\mathbf{Q}_2$ that has a projective representation that admits a lifting of conductor 2^5 and determinant μ (this just involves checking the possible twists of the representations of lower minimal conductor). Thus the local behavior of K and K' at 2 is completely determined and $K_2 \cong K'_2$.

At 5 the fields K and K' must have liftings of conductor 2 and determinant ν. Again this uniquely specifies the local representation so $K_5 \cong K'_5$.

Since the completions of K and K' are isomorphic for p = 2,3,5, and 11, each prime of K over 2,3,5, or 11 must split completely in L/K.

Since L/K is unramified we must have $D_{L/\mathbb{Q}} = (D_{K/\mathbb{Q}})^{60}$. There are 5 primes above 2 in K, each of which contributes 2^{18} to $D_{K/\mathbb{Q}}$ (by looking at the ramification groups one sees that the valuation of the different is 6; the residue class degree is 3). There are 12 primes above 5 in $\mathbb{Q}$, each of which contributes 5^8 to the discriminant. Thus

$$D_{K/\mathbb{Q}} = 2^{90}5^{96}.$$

and

$$(1/3600)\log D_{L/\mathbb{Q}} = (1/60)\log D_{K/\mathbb{Q}} = (3/2)\log(2)+(8/5)\log(5).$$

If we insert this in the inequality *) above, bring all of the constants to the left, and do some arithmetic we get

$$.5368 \geq (2/3600)\ \Sigma\, C(m,P).$$

If this inequality is true then it must be true if we sum only over those primes P of L that lie over 2,3,5, and 11. Since any such prime splits completely in L/K we see that if the above inequality is true then

$$.5368 \geq (2/60)\ \Sigma\, C(m,P)$$

where P now runs through the primes of K above 2,3,5 and 11. This sum can be calculated in a straightforward way: there are 5 primes of K above 2 each with norm 8, 20 primes of K above 3 each of norm 27, 12 primes of K above 5 each of norm 5, and 30 primes of K above 11 each of norm 121. If we total these contributions we get about .5948.. so we have contradicted inequality *). The only recourse is to have K = K'. By the remarks above this finishes the proof of the proposition and thereby also the proof of Theorem 10. q.e.d.

Remarks: 1) The machine verification of the hypothesis that f' is an eigenform for T_{11} is a straightforward exercise; it is a triviality when compared with the computations required for the "Fact" above. One needs to know about 1000 coefficients of f'; these are easy to produce since we have expressed f' as a quotient of known forms. The only reason that this computation has not been done is that the proof of Theorem 10 was obtained recently. At any rate it would take a hard-hearted cynic to disbelieve this hypothesis.

2) The original stimulus for the theorem above was a conversation with Gunter Harder, who suggested that something like the proposition above might be provable.

3) The conclusions of Theorem 10 imply that the vector space V contains a unique newform with the given eigenvalues at 3 and 11. Similarly there is another newform that can be obtained from the first by an automorphism of **C** that takes i to -i and fixes the fifth roots of unity (this changes the T_3 eigenvalue). One actually wants to be able to completely describe V, without regard to the T_3 and T_{11} eigenvalues. This could be accomplished by improving the discriminant bounds in such a way as to be able to drop the contributions from primes above 3 and 11. Certainly the bounds obtained under the assumption of the Generalized Riemann Hypothesis are sufficient to obtain the desired result: dim V = 14; V contains a 12-dimensional space of oldforms and two distinct icosahedral newforms.

BIBLIOGRAPHY

[Aho-Hopcroft-Ullman]
A. Aho, J. Hopcroft, and J. Ullman: The Design and Analysis of Computer Algorithms. Addison Wesley Pub. Co., 1974.

[Antwerp IV]
Modular Functions of One Variable IV. Lecture Notes in Mathematics 476, Springer Verlag, 1975.

[Artin]
E. Artin: Zur Theorie der L-Reihen mit allgemeinen Gruppencharakteren. Hamb. Abh. 8 (1930), p. 292-306. (Collected Works p. 165-179)

[Artin-Tate]
E. Artin and J. Tate: Class Field Theory. W. A. Benjamin, 1967.

[Atkin-Lehner]
A. O. L. Atkin and J. Lehner: Hecke Operators on $\Gamma_0(m)$. Math. Ann. 185 (1970), p. 134.

[Bareiss]
E. Bareiss: Computational Solutions of Matrix Problems over an Integral Domain. J. Inst Maths. Applics. 10(1972), p. 68.

[Borel]
A. Borel: Formes automorphes et séries de Dirichlet. Séminaire N. Bourbaki 466, 1975.

[Cassels-Fröhlich]
J. W. S. Cassels and A. Fröhlich: Algebraic Number Theory. Academic Press, 1967.

[Cayley]
A. Cayley: Collected Mathematical Papers. Cambridge University Press, 1891.

[Deligne]
P. Deligne: Les Constantes des Equations Fonctionnelles des Fonctions L. Lecture Notes in Mathematics 349, p. 501-597, Springer Verlag, 1973.

[Deligne-Serre]
P. Deligne and J.-P. Serre: Formes modulaires de poids 1. Ann. Scient. Ec. Norm. Sup. (4), 7(1974), p.507-530.

[Dixon]
J. Dixon: The Structure of Linear Groups. Van Nostrand, 1971.

[Gérardin]
P. Gérardin: Construction de Séries discrètes p-adiques. Lecture Notes in Mathematics 462, Springer Verlag, 1975.

[Hecke]
E. Hecke: Mathematische Werke. Vandenhoeck und Ruprecht, 1970.

[Hunter]
J. Hunter: The minimum discriminants of quintic fields. Proc. Glasgow Math. Assoc. 3 (1957) p. 57-67.

[Huppert]
B. Huppert: Endliche Gruppen I. Springer Verlag, 1967.

[Jacquet-Langlands]
H. Jacquet and R. P. Langlands: Automorphic Forms on GL(2). Lecture Notes in Mathematics 114, Springer Verlag, 1970.

[Knuth]
D. Knuth: The art of computer programming, vol. 2, Seminumerical Algorithms. Addison-Wesley, 1969.

[Koch, CP]
H. Koch: Classification of the primitive representations of the Galois group of local fields (preprint).

[Koch, DP]
H. Koch: Die irreduziblen Darstellungen von Primzahlgrad der Galoisschen Gruppe eines lokalen Körpers (preprint).

[Langlands]
R. P. Langlands: Base Change for GL(2). Notes for Lectures at the Institute for Advanced Study, 1975.

[Li]
W. Li: Newforms and Functional Equations. Math. Ann. 212(1975), p. 285-315.

[Miller]
V. Miller: Diophantine and p-adic Analysis of Elliptic Curves and Modular Forms. thesis, Harvard University, 1975.

[Odlyzko]
A. Odlyzko: Bounds on Discriminants. Unpublished manuscript, Bell Labs, November, 1976.

[Ogg, CL]
A. Ogg: On a convolution of L-series. Inv. Math. 7(269), p. 297-312, 1969.

[Ogg, MF]
A. Ogg: Modular Forms and Dirichlet Series . W. A. Benjamin, 1969.

[Ogg, HC]
A. Ogg: Hyperelliptic Modular Curves. Bull. de la Soc. Math. de France, 102(1974), p. 449-462.

[Ogg, RP]
A. Ogg: Rational Points on Certain Elliptic Modular Curves. AMS Symposia in Pure Math vol. XXIV, p. 221-231, 1973.

[Poitou]
G. Poitou: Minorations de discriminants. Sém. Bourbaki 479, 1976.

[Rankin]
R. Rankin: Contributions to the theory of Ramanujan's function $\tau(n)$ and similar arithmetical functions II. Proc. Camb. Phil. Soc 35(1939), 357-372.

[Serre, CL]
J.-P. Serre: Corps Locaux. Hermann, 1968.

[Serre, D]
J.-P. Serre: Modular Forms of Weight One and Galois Representatons. Proceedings of a symposium at the University of Durham, Academic Press, 1977.

[Serre, RG]
J.-P. Serre: Representations Lineaires des Groupes Finis. Hermann, 1967.

[Shimura]
G. Shimura: On the holomorphy of certain Dirichlet series. Proc. London Math. Soc. 31(1975), p. 79.

[Suprenenko]
P. Suprenenko: Matrix Groups. AMS translations, 1976.

[Tate, N133]
J. Tate: various notes on the work of Tate, Atkin, Flath, Kottwitz, Tunnell, and Weisinger on the tetrahedral representation of conductor 133 (unpublished).

[Tate, NFZ]
J. Tate: some notes on fractional compulsory zeroes of modular forms with a character (unpublished).

[Thompson-Ritchie]
K. Thompson and D. Ritchie: The UNIX Operating System. Comm. ACM 17(1974), p. 365.

[Weil, ED]
A. Weil: Exercices Dyadiques. Inventiones, 27 (1974), p.1-22.

[Weil, CF]
A. Weil: Sur la théorie du corps de classes J.Math. Soc. Japan, 3 (1951), p. 1-35.

[Weil, OU]
A. Weil: Sur certains groupes d'opérateurs unitaires. Acta Math. 111(1964), p. 143-211.

[Zimmer]
H. Zimmer: Computational Problems, Methods, and Results in Algebriac Number Theory. Lecture Notes in Mathematics 262, Springer Verlag, 1972.

[Zassenhaus]
H. Zassenhaus: On embedding of an order into a maximal order. Lecture Notes in Mathematics 353, p. 204-221, 1973.

APPENDIX 1: The Sextic Resolvent

Let $f(x) = x^5 + bx^4 + cx^3 + dx^2 + ex + f = \prod (x-x_i) \in F[x]$ be a monic quintic polynomial with square discriminant D^2 where

$$D = \prod_{1 \leq i < j \leq 5} (x_i - x_j) \quad , \; D \in F.$$

It is left to the reader to verify that the discriminant is given in terms of the coefficients b,c,d,e,f as

$$D^2 = 3125f^4 + A_3 f^3 + A_2 f^2 + A_1 f + A_0$$

where

$$A_3 = 2500be + 2000b^3d - 3750cd + 256b^5 + 2250bc^2 - 1600b^3c$$

$$\begin{aligned} A_2 = {} & (-50b^2 + 2000c)e^2 + (2250d^2 + (160b^3 - 2050)d + (1020b^2c^2 - 192b^4c - 900c^3))e \\ & + (-900bd^3 + (560b^2c - 128b^4 + 825c^2)d^2 + (144b^3c^2 - 630bc^3)d + (108b^5 - 27b^2c^4)) \end{aligned}$$

$$\begin{aligned} A_1 = {} & (-1600d + 160bc - 36b^3)e^3 + \\ & (1020bd^2 + (144b^4 - 74bc^2 + 560c^2)d + 24bc^3 - 6b^3c^2)e^2 + \\ & ((24b^2 - 630c)d^3 + (356bc^2 - 80b^3c)d^2 + (18b^2c^3 - 72c^4)d)e + \\ & (108d^5 + (16b^3 - 72bc)d^4 + (16b^3 - 4b^2c^2)d^3)) \end{aligned}$$

$$\begin{aligned} A_0 = {} & 256e^5 + \\ & (-192bc + 144b^2c - 128c^2 - 27b^4)e^4 + \\ & ((144c - 6b^2)d^3 + (18b^3c - 80bc^2)d + 16c^4 - 4b^2c^3)e^3 + \\ & (-27d^4 + (18bc - 4b^3)d^3 + (b^2c^2 - 4c^3)d^2)e^2. \end{aligned}$$

If $y = y(x_1, x_2, x_3, x_4, x_5)$ is invariant under precisely the dihedral group $D_5 \subset S_5$ of permutations of the x_i (generated by (12345) and (15)(24)), then the equation satisfied by y is called a sextic resolvent of f(x). The function $y = x_1x_2 + x_2x_3 + x_3x_4 + x_4x_5 + x_5x_1$ is a possibility, but it is more convenient to choose

$$y_1 = x_1x_2 + x_2x_3 + x_3x_4 + x_4x_5 + x_5x_1 - x_1x_3 - x_2x_4 - x_3x_5 - x_4x_1 - x_5x_2.$$

Here we define the sextic resolvent of the quintic f(x) to be the equation

$$g(y) = y^6 + a_1y^5 + a_2y^4 + a_3y^3 + a_4y^2 + a_5y + a_6 = \Pi(y - y_i)$$

satisfied by the six conjugates of y_1. The coefficients of the sextic are given by

$$a_1 = a_3 = 0 \qquad a_5 = 32D$$

$$a_2 = 8bd - 3c^2 - 20e$$

$$\begin{aligned} a_4 = {} & -64b^3f + 16b^2d^2 + 16b^2ce - 16bcd^2 + 240bcf \\ & -112bde + 3c^4 - 8c^2e + 16cd^2 - 400df \end{aligned}$$

$$\begin{aligned} a_6 = {} & -c^6 + c^4(8bd + 28e) + c^3(-16b^2e - 16d^2 + 48bf) \\ & + c^2(-16b^2d^2 - 112bde - 80df - 176e^2) \\ & + c(64b^3de - 192b^2df + 64bd^3 + 224b^2e^2 - 640bef + 224d^2e + 4000f^2) \\ & -64b^4e^2 + 384b^3ef - 128b^2d^2e - 1600b^2f^2 + 640bd^2f \\ & -64bde^2 - 64d^4 - 1600def + 320e^3. \end{aligned}$$

Remark: These formulae can essentially be found in [Cayley, vol. IV, p.318] (his discriminant is not quite the same as the discriminant D^2 above). All later sources that I consulted lacked the courage to display the above equations; they only have the sextic resolvents of quintics of the form $x^5 + ax + b$.

The industrious reader can now express the discriminant of the resolvent sextic in terms of the coefficients b,c,d,e,f of the original quintic. One interesting fact is that, unlike the cubic resolvent of a quartic, the discriminant of the quintic does not necessarily divide the discriminant of the sextic. Thus, despite the fact that the same primes ramify in the root fields of an A_5 quintic and its sextic resolvent, "spurious" primes can divide the discriminant of the quintic and not divide the discriminant of the sextic (or vice versa).

APPENDIX 2: Extensions of $\mathbb{Q}_5$ of degree 5

Let v_5 be the valuation on $\mathbb{Q}_5$ with $v_5(\mathbb{Q}_5{}^*) = \mathbb{Z}$.

Proposition: Let $F/\mathbb{Q}_5$ be a ramified extension of degree 5. Choose an Eisenstein polynomial

$$f(x) = x^5+a_1x^4+a_2x^3+a_3x^2+a_4x+a_5 \in \mathbb{Z}_5[x]$$

whose root field is F. Assume that $v_5(a_1) = 1$. Then F is is a galois extension of $\mathbb{Q}_5$ if and only if the following conditions are satisfed:

$$v_5(a_4)\geq 2, \quad v_5(a_3)\geq 2, \quad v_5(a_2)\geq 2, \quad v_5(a_1+a_5)\geq 2 \quad .$$

Remark: We can always insure that $v_5(a_1) = 1$ by replacing f(x) by f(x+5) if necessary.

Proof: Let r be a root of f(x) in F, so that r is a uniformizing parameter of F. Then $F/\mathbb{Q}_5$ is galois if and only if all other roots of f(x) are also in F.

$$\begin{aligned} f(r+xr^2) &= f(r)+ xr^2f'(r)+ x^2r^4f^{(2)}(r)/2! + x^3r^6f^{(3)}(r)/3! \\ &\quad + x^4r^8f^{(4)}(r)/4! + x^5r^{10}f^{(5)}(r)/5! \\ &= xr^2(5r^4+4a_1r^3+3a_2r^2+2a_3r+a_4) \\ &\quad +x^2r^4(10r^3+6a_1r^2+3a_2r+a_3) \\ &\quad +x^3r^6(10r^2+4a_1r+a_2) \\ &\quad +x^4r^8(5r+a_1) \\ &\quad +x^5r^{10}. \end{aligned}$$

If $v_5(a_4) = 1$ then the above equation cannot have a nonzero x as a solution since then the valuation of xr^2a_4 would be strictly smaller than the valuation of any other term. Similarly for a_3 and a_2 so that the first three conditions above are certainly necessary if f(x) is to have 5 roots in F. If we assume the first three conditions then modulo r^{11} the above equation is

$$f(r+xr^2) \equiv xr^2(4a_1r^3+x^4r^8) \bmod r^{11}.$$

If this is to have a nonzero solution then we must have

$$4(a_1/r^5)+ x^4 \equiv 0 \bmod r.$$

which is the same as

$$a_1/r^5 \equiv 1 \bmod r$$

since the 4-th power of any nonzero element of the residue field is 1. The equation $f(r) = 0$ says that $r^5+a_5 \equiv 0 \bmod r^6$ and the above equation is

$$a_1 \equiv -a_5 \bmod r^6 \quad \text{or} \quad v_5(a_1+a_5) \geq 2$$

which shows the necessity of the above 4 conditions.

Assume the validity of these 4 conditions. Define

$$g(x) = f(r+xr^2)/xr^{10} \in \mathbb{Z}_5[r][x].$$

The equations above show that $g(x) \equiv 0 \bmod r$ if x is any unit. The equations above also show that

$$g'(x) = 4x^3 + r(.....)$$

so that, by Hensel's Lemma, $g(x)$ has 4 roots in F. q.e.d.

Remark: It is curious that the fancy form of Hensel's Lemma does not guarantee that $r+xr^2$ can be refined to a root of $f(x)$; in order to deduce this one must form the function $g(x)$ and apply the simple version of Hensel's Lemma.

Examples:

The following two polynomials both arose in the search for A_5 fields; each of them has splitting field A_5, is unramified outside 2 and 5, and has A_4 as its decomposition group at 2.

1) $f(x) = x^5+10x^3-10x^2+35x-18$.

Since $v_5(f(-2)) = 1$, we consider the equation for $y = x+2$ which is $g(y) = y^5-10y^4+50y^3-150y^2+275y-190$. By the above proposition the splitting field of $f(x)$ over $\mathbf{Q}_5$ must be cyclic of order 5.

2) $f(x) = x^5+10x^4-10x^3-20x^2+30x+32$.

Since $v_5(f(-2)) = 2$ it is natural to take $y = 5/(x+2)^2$ as a uniformizing parameter; the equation for y is $g(y) = 16y^5-180y^4+600y^3-400y^2+100y-5$, so that by the proposition above the splitting field of $f(x)$ is dihedral at 5!

APPENDIX 3: The A_4 Extension of $\mathbb{Q}_2$

Let K be the unique A_4 extension of $\mathbb{Q}_2$. It is easy to verify that K is the splitting field of

$$f(x) = x^4 - 2x^3 - 2x^2 + 2 = 0$$

(see Table 3.3). Put $G = G(K/\mathbb{Q}_2)$ and let $\tau \in G$ be an element of order 2. Let E be the fixed field of τ so that K is a quadratic extension of E; let $\Theta: E^* \longrightarrow \pm 1$ be the norm residue mapping determined by K/E.

According to the theory of chapter 2 there is a character $\chi: K^* \longrightarrow \mathbf{C}^*$ of conductor 4 that is centric for the unique faithful projective representation $T: G \longrightarrow PGL_2(\mathbf{C})$. Also there for such a χ there are two characters $\psi: E^* \longrightarrow \mathbf{C}^*$ of conductor 3 such that

$$\chi = \psi \circ N_{K/E}.$$

One of these characters can be obtained from the other by multiplication by Θ. The object of this appendix is to give a formula for such a ψ.

In order to do this explicitly it is necessary to introduce some notation describing K, E, and the action of G. Let F be the fixed field of the klein group of order 4 in $G \cong A_4$. The field F is the unramified cubic extension of $\mathbb{Q}_2$ and can be obtained by adjoining a primitive 7-th root of unity to $\mathbb{Q}_2$.

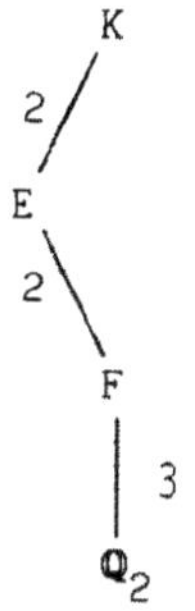

Let u be a root of $f(x) = 0$, and let $v = N_{K/E}u$. Since $f(x)$ is an eisenstein polynomial u is a uniformizing parameter for K and v is a uniformizing parameter for E.

Let z be the unique 7-th root of unity in K such that

$$u \equiv u + zu^2 \mod u^3.$$

It is easy to use the equation $f(u^\tau) = 0$ to check that z is well defined (i.e., that $z \neq 0$) and that $Tr(z) = 0$, where we let $Tr(x)$ denote

$$Tr_{\mathbf{F}_8/\mathbf{F}_2}\bar{x} \quad \in \{0,1\}$$

for x in K and $\bar{x}$ its image in the residue field.

Let $\varphi \in G$ be the unique representaive of frobenius that fixes u. Then G is generated by φ and τ and $K = \mathbf{Q}_2(u,z)$. Moreover

$$z^\tau = z \quad z^\varphi = z^2 \quad u^\varphi = u$$

so that the action of G on K will be completely specified if we can compute u^τ.

Using the equation $f(u) = 0$ together with some iterative techniques it is possible to compute 2 as a power series in u:

1) $$2 = u^4 + u^6 + u^7 + \ldots$$

Using this expression together with Hensel's Lemma (applied to $f(u^\tau) = 0$) we get

2) $$u^\tau = u + zu^2 + z^2u^2 + z^5u^3 \ldots$$

Note that

$$v = N_{E/F} = u\,u^{\tau} = u^2 + zu^3 + z^2u^4 + \ldots$$

so that be combining 1) and 2) we get

3) $$2 = v^2 + z^6v^3 \;. \;.$$

Let σ be an element of G that is of order 2 and distinct from τ. Then τ induces an automorphism of E (since σ and τ commute) and in fact G(E/F) is generated by τ. Since the roots of f(x) are the four conjugates of u under the klein group and the constant term of f(x) is 2 we see that

$$2 = N_{K/F}u = N_{E/F}v = v^{1+\sigma}$$

so

$$v^{\sigma} = 2/v = v + z^6v^2 + z^4v^3 \;. \;. \;.$$

Suppose that a character $\psi : E^* \longrightarrow \mathbf{C}^*$ satisfies

A) $\psi^{\sigma-1} = \Theta$

B) $\psi^{\varphi} = \psi$

C) $a(\psi) = 3.$

Then by the theory of chapter 2 (specifically the proof of Proposition 3), A) implies that Ind(ψ) is a linear representation of G_F that lifts the restriction $T|G(K/F)$. The assertion B says that Ind(ψ) is invariant under $G(F/\mathbf{Q}_2)$ and therefore Ind(ψ) can be extended to a representation of G that lifts the projective representation T. The theory of Chapter 2 then shows that C) implies that the resulting representation has conductor 5, which is the smallest conductor of a representation that lifts T. Thus we wish to find a ψ that satisfies A, B, and C.

Note that by direct calculation (or by [Serre, CL, prop. 5, p. 236]) it follows that

$$\Theta(U_E^2) = \Theta(z^i) = \Theta(v) = 1$$

$$\Theta(1+av) = (-1)^{Tr(az^5)}.$$

Thus if ψ satisfies A and C then

$$\psi(1+bv^2+. \ .) = \psi((1+bzv)^{\sigma-1}) = \Theta(1+bzv) = (-1)^{Tr(bz^6)}$$

and

$$1 = \Theta(v) = \psi(v^{\sigma-1}) = \psi(1+z^6v+z^4v) = -\psi(1+z^6v).$$

Suppose that ψ satisfies A, B, and C and put

$$Y_2(\bar{b}) = \psi(1+bv^2) \qquad Y_1(\bar{a}) = \psi(1+av).$$

Then Y_2 is a character of the additive group of the residue field $\mathbf{F}_8$ and, since ψ is a homomorphism,

$$Y_1(a+a')Y_2(aa') = Y_1(a)Y_1(a')$$

so that Y_1 is a quadratic character of $\mathbf{F}_8$ (in the sense of [Weil, OU]) with respect to the character Y_2.

The character Y_2 is uniquely determined and is given above. The general theory of quadratic characters over finite fields of characteristic 2 is messy. An explicit construction is given in [Gerardin]; using this construction it follows that

$$Y_1(a) = i^{Tr(az^3)}(-1)^{Tr(a^3z^2)}$$

is a quadratic character with respect to Y_2 and that it satisfies the necessary condition

$$1 = \psi(v^{\sigma-1}) = \psi(1+z^6v+z^4v^2) = Y_1(z^6)Y_2(z^4).$$

Theorem: Define $\psi: E^* \longrightarrow \mathbb{C}^*$ by

$$\psi(v^m z^n(1+av+bv^2+\ .\ .\ .\ )) = Y_1(\bar{a})Y_2(\bar{b})$$

where Y_1 and Y_2 are defined above. Then ψ is a character of E^* that satisfies the conditions A, B, and C above.

Proof: The assertions B and C are obvious, and A follows by the preceding remarks. The fact that ψ is a homomorphism follows from the fact that Y_1 is a quadratic character with respect to Y_2. q.e.d.

In chapter 4 the above formula is used to give the local component of a global idele class character ψ that is defined on a sextic extension of $\mathbb{Q}$.

APPENDIX 4:

S_4 extensions of $\mathbb{Q}$ unramified outside 2 and 5

In order to make a complete description of the space of modular forms of weight one and level 800 it is necessary to describe the modular forms coming from S_4 extensions of $\mathbb{Q}$.

Theorem: There are exactly three S_4 extensions of $\mathbb{Q}$ that are unramified outside of 2 and 5. The minimal conductors of liftings of the associated projective representations are $2^9 5^2, 2^9 5^2$, and $2^8 5^2$.

Proof: Consider the following table of 3 quartic polynomials, together with their discriminants and factorizations mod 3,7 and 13:

f(x)	discr_f	mod 3	mod 7	mod 13
$x^4+4x^3+2x^2-2$	$-2^9 5^2$	1+3	4	4
x^4+4x^2-8x+2	$-2^{11} 5^2$	1+3	4	1+1+2
$x^4-4x^2+16x-18$	$-2^{11} 5^4$	1+3	1+1+2	4

(here 1+3 means that f(x) factors as a linear times a cubic, etc.)

The information in this table implies that these three polynomials have non-ismorphic splitting fields, each of which is an S_4 extension of $\mathbb{Q}$ unramified outside 2 and 5.

At 5 each of the above S_4 fields has S_3 as its decomposition group. Hence the minimal conductor at 5 is as claimed in the theorem.

At 2 the decomposition group is the dihedral group of order 8. This can be verified by showing that the resolvent cubic has one root in $\mathbb{Q}_2$ and showing that the splitting field of f(x) over $\mathbb{Q}_2$ has more than one quadratic subextension. The minimal conductor is computed in the table at the end of this appendix by using proposition 3 in chapter 2. The following fact is used: if E_1 and E_2 are quadratic extensions of a local field F with $b_0(E_1/F) < b_0(E_2/F)$ and $K = E_1E_2$ then

$$b_0(K/E_2) = b_0(E_1/F)$$

$$b_0(K/E_1) = 2b_0(E_2/F)-b_0(E_1/F) \quad \text{if } E_1/F \text{ is ramified}$$

$$= b_0(E_2/F) \quad \text{otherwise}$$

Finally we must show that the above three extensions are the only S_4 extensions of $\mathbf{Q}$ that are unramified outside 2 and 5. Let $L/\mathbf{Q}$ be such an extension. Define subfields of L by making them correspond to subgroups of S_4 as follows:

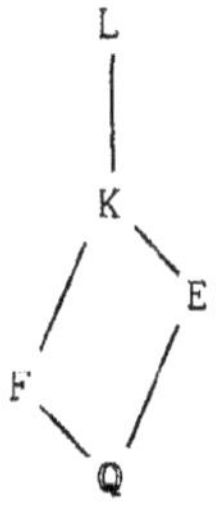

$G(L/\mathbf{Q}) \cong S_4$

$G(L/K) \cong V$

$G(K/\mathbf{Q}) \cong S_3$

$G(L/E) \cong D$

$G(L/F) \cong A_4$

where V is the Klein group of order 4 and D is some fixed dihedral group of order 8 (= a 2-sylow subgroup of S_4).

The field F must be one of the 7 quadratic fields unramified outside 2 and 5: $\mathbf{Q}(\sqrt{-1}),\mathbf{Q}(\sqrt{-2}),\mathbf{Q}(\sqrt{2}),\mathbf{Q}(\sqrt{-5}),\mathbf{Q}(\sqrt{5}),\mathbf{Q}(\sqrt{-10}),\mathbf{Q}(\sqrt{10})$. Since none of the class numbers of these fields is divisible by 3, K/F must give a tame local extension of degree 3 at a prime above 2 or 5. This can happen only if the cardinality of the residue field is congruent to 1 mod 3, i.e. either 2 or 5 is inert in F. Thus $F = \mathbf{Q}(\sqrt{-2})$, $\mathbf{Q}(\sqrt{2})$, or $\mathbf{Q}(\sqrt{5})$. The unit $1+\sqrt{2}$ (resp. $1+\sqrt{5})/2$) in $\mathbf{Q}(\sqrt{2})$ (resp. $\mathbf{Q}(\sqrt{5})$) generates the residue field $\mathbf{F}_{25}$ (resp. $\mathbf{F}_4$) at 5 (resp. 2). In these cases there is no idele class character of F unramified outside the primes above 2 and 5 so the required K/F does not exist. So from now on $F = \mathbf{Q}(\sqrt{-2})$. In this case U_{F_5} contains a unique subgroup of index 3 so that K is unique. K is the splitting field of the cubic

$$g(x) = x^3 - x^2 + 2x + 2; \qquad \mathrm{discr}_g = -200 = -2^3 5^2.$$

Let η be the root of g(x) with $E = \mathbf{Q}(\eta)$. Then

$$g(x) \equiv (x-53)(x^2+52x-58) \bmod 256$$

and in the ring of integers of E we have

$$2 = \mathfrak{P}_1 \mathfrak{P}_2^2 \qquad \eta \equiv 1 \bmod \mathfrak{P}_1 \qquad \eta \equiv 0 \bmod \mathfrak{P}_2$$

$$N(\eta) = -2 \qquad N(\eta+1) = 2 \qquad N(\eta-1) = -4$$

$$(\eta) = \mathfrak{P}_2 \qquad (\eta+1) = \mathfrak{P}_1 \qquad (\eta-1) = \mathfrak{P}_1^2$$

so $\eta^2-\eta-1 = -(\eta+1)^2/(\eta-1)$ is a unit. The Minkowski bound for E is

$$(4/\pi)(6/27)\sqrt{200} < 5$$

so that the class number of E is one ($\mathfrak{P}_1$ and $\mathfrak{P}_2$ are principal by the norms given above and 3 is inert).

In order to make an S_4 extension of $\mathbb{Q}$ we need a Klein group extension L/K. By class field theory L corresponds to a subgroup

$$N \subset C_K, \qquad G(L/K) \cong C_K/N \cong V.$$

By Shafarevic's theorem a subgroup N gives an S extension if and only if

a) the action of $G(K/\mathbb{Q}) \cong S_3$ on $V \cong C_K/N$ is the usual action of $S_3 \cong S_4/V$ on V.

b) the fundamental class $\xi(K/\mathbb{Q}) \in H^2(G(K/\mathbb{Q}),C_K)$ goes to the class determined by S_4 in $H^2(S_3,V)$ via the reciprocity map.

The action of S_3 on V corresponds to the irreducible 2-dimensional representation of S_3 over $\mathbb{F}_2$. For this representation $H^2(S_3,V)= 0$ (e.g. S_4 is a semidirect product) so condition b) is automatically satisfied.

Thus the S_4 extension L corresponds to a nontrivial S_3 homomorphism from C_K to V. In order to describe these more concretely let t be a 3-cycle in $G(L/\mathbb{Q})$ and let s be a transposition (= element of order 2 not in G(L/K)) that fixes E. Let X be the unique nontrivial character of G(L/K) that is fixed by s. The other two non-trivial characters of G(L/K) are χ^t and χ^{t^2}. Thus χ corresponds to a character (also written χ) of C_K:

$$\chi : C_K \longrightarrow \pm 1,$$

with the following properties:

1) $\chi^s = \chi$
2) $\chi^N = 1$, where $N = 1+t+t^2$.

Conversely any χ satisfying these two properties uniquely determines a nontrivial S_3 homomorphism from C_K to V and therefore an S_4 extension of $\mathbf{Q}$.

Property 1) implies that there is a character $\psi : C_E \longrightarrow \mathbf{C}^*$ with

$$*) \quad \chi = \psi \circ N_{K/E}.$$

In fact there are exactly two such characters ψ; if $\Theta : C_E \longrightarrow \pm 1$ is the reciprocity mapping for K/E then $\Theta\psi$ is another character satisfying *. Also ψ is of order 2 since it is a can be interpreted as a character of $G(L/E) \cong D$.

In order for any character $\psi : C_E \longrightarrow \pm 1$ to determine a χ and hence an S_4 extension it suffices to have

$$\psi(N_{K/E}x) = 1 \text{ if } x = y^{1+t+t^2} = N_{K/F}y \text{ for any } y \in C_K.$$

Since

$$N_{K/E}N_{K/F}C_K = N_{F/\mathbf{Q}}N_{K/F}C_K = N_{K/\mathbf{Q}}C_K = N_{F/\mathbf{Q}}C_K$$

this can be interpreted as saying that the map

$$N_{F/\mathbf{Q}}C_F \longrightarrow C_{\mathbf{Q}} \longrightarrow C_E \longrightarrow \pm 1$$

is trivial.

Note that there is one real place, say ∞ , of E and that this place ramifies in K/E; i.e. Θ_∞ is non-trivial. Thus the two ψ's corresponding to χ can be distinguished by their component at ∞ ; we could fix ψ by requiring that $\psi_\infty = 1$.

Also note that the decomposition group of 5 in $G(K/\mathbb{Q})$ is all of $G(K/\mathbb{Q}) \cong S_3$. Therefore the decomposition group of 5 in all of S_4 is S_3 and the unique prime above 5 in K must split completely in L/K. Since K/E is unramified at 5, ψ is unramified at 5.

The above remarks reduce the problem of finding S_4 extensions of $\mathbb{Q}$ unramified outside 2 and 5 to the problem of finding the ψ's. More precisely:

There is a one to one correspondence between S_4 extensions of $\mathbb{Q}$ unramified outside 2 and 5 and characters $\psi: C_E \longrightarrow \pm 1$ unramified outside 2 (in particular, unramified at ∞) such that

$$\psi(N_{F/\mathbb{Q}} C_F) = 1.$$

Because the class number of E is 1, a ψ is determined by its components on the units at $\mathfrak{P}_1$ and $\mathfrak{P}_2$. Note that $E_{\mathfrak{P}_1} \cong \mathbb{Q}_2$ and $E_{\mathfrak{P}_2} \cong F_2$. Thus ψ is uniquely determined by a character on the group

$$U_{\mathbb{Q}_2}/(U_{\mathbb{Q}_2})^2 \times U_{F_2}/(U_{F_2})^2.$$

This group is an elementary abelian group of order 2^5.

Conversely any character on this group determines a ψ if it is trivial on the image of the units and on the image of $N_{F/\mathbb{Q}} C_F$. It is now a straightforward exercise to verify that -1, $\eta^2 - \eta - 1$, (which are both units) and 3 ($\in N_{F/\mathbb{Q}} C_F$) are nontrivial and independent so that ψ is actually a character of an elementary abelian group of order (at most) 2^2. This finishes the proof of the above theorem

Remark: The proof actually gives a construction of the character ψ and as a consistency check it can be used to verify that the resulting three extensions are associated to the original three quartic polynomials. The component of ψ at $\mathfrak{P}_2$ gives the extension K''/E'' in the diagram in Table App4.1; the above construction produces characters of conductor 4,5, and 5 which checks with the values tabulated in Table App4.1.

Table App4.1: Some Dihedral Extensions of $\mathbf{Q}_2$

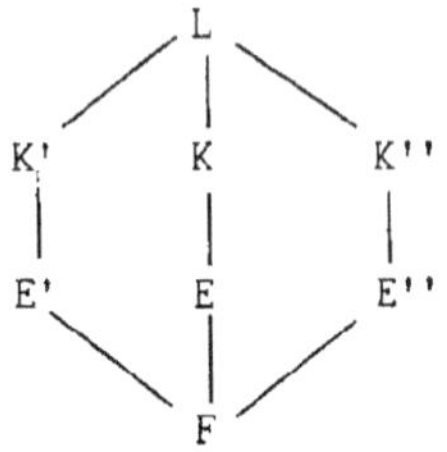

$F = \mathbf{Q}_2$; $f(x) = x^4-ax^3+bx^2-cx+d = \Pi(x-x_i)$; $K' = F(x_1)$

L = the splitting field of F; $E'' = F(\sqrt{\mathrm{discr}_f})$

$g(y) = y^3-by^2+(ac-4d)y+(4bd-a^2d-c^2) = \Pi(y-y_i)$

If L is dihedral over F then the "cubic resolvent" g(y) must have one root in F;

say $y_1 = x_1x_2+x_3x_4 \in F$.

If $z = x_1x_2$ then $E' = F(z)$, $z^2-y_1x+d = 0$

$\mathrm{discr}_f = \mathrm{discr}_g = D_{K'/F} = (N_{E'/F}D_{K'/E'})(D_{E'/F})^2$

f(x)	g(y)	y_1 mod 256	E'	E
$x^4+4x^3+2x^2-2$	$y^3-2y^2+8y+16$	-22	F(3)	$F(\sqrt{10})$
x^4+4x^2-8x+2	$y^3-4y^2-8y-32$	12	F(2)	$F(\sqrt{-1})$
$x^4-4x^2+16x-18$	$y^3+4y^2+72y+32$	92	F(6)	$F(\sqrt{5})$

$v_2(\mathrm{discr}_f)$	$b_0(K'/E')$	$b_0(L/K)$	$b_0(K''/E'')$	$b_0(K/E)$	r	r'	cond(f)
9	4	5	3	1	2	4	9
11	4	7	4	3	1	6	9
11	4	r	4	2	0	4	8

r and r' are as in the proposition 3 in chapter 2; cond(f) is the minimal conductor of a lifting of the projective representation associated to the quartic polynomial and is computed via that proposition.

Appendix 5

Some of the interesting aspects of this thesis involve the choice of algorithms and implementations used to obtain the "numerical" results; e.g. the integral closure algorithm and the algorithms used to compute the global ψ on the sextic field. The lack of any detailed discussion of these algorithms has the effect of concealing a whole panoply of ideas and techniques. This appendix is an attempt to slightly atone for this by briefly discussing two techniques used in some of the programs used: 1) the representation and manipulation of large integers, and 2) the exact of large linear systems of equations whose coefficients lie in a number field. These examples are chosen because of their mathematical and algorithmic simplicity, but hopefullly they will impart some of the flavor of the sort of problems encountered when one wants to use a machine for serious computations.

Throughout this appendix let $p_1 < p_2 < \ldots < p_m$ be a fixed sequence of rational primes, and let

$$P_i = p_1 p_2 \cdots p_i .$$

In practice the p_i are taken to be roughly equal to the word size and m is chosen so that P_m is bigger than any integer that might arise in a specific problem at hand.

Large Integers

Since the integers that arise in number theoretic computations are frequently larger than the word size of a specific computer (indeed, frequently larger than the word size of any existing computer) we are interested in describing some data representations for large integers.

Let $x \in \mathbb{Z}$. One approximate multiword representation

$$x \longrightarrow [x_1, x_2, \ldots, x_n], \quad 0 \leq x_i < p_i,$$

is the so-called mixed base representation, where the x_i are uniquely determined by

$$x = x_1 + x_2P_1 + x_3P_2 + \ldots + x_nP_{n-1} \bmod P_n$$

$$0 \leq x_i < p_i.$$

If we let n go to infinity then

$$\text{if } x > 0 \text{ then } x_i = 0 \quad \text{for sufficiently large } i$$

$$\text{if } x < 0 \text{ then } x_i = p_i - 1 \quad \text{for sufficiently large } i.$$

The mixed-base representation is related to the usual decimal expansion of x (which could be obtained by replacing all of the p_i with 10) or the more usual representation used for multiprecision routines on computers (which could be obtained by replacing all of the p_i by the word size of the machine).

Another approximate multiword representation is the modular representation whereby

$$x \longrightarrow [x_1, \ldots, x_n], \qquad x \equiv x_i \bmod p_i, \quad 0 \leq x_i < p_i.$$

If we fix x to lie in any interval of length less than P_n then x is uniquely determined by $[x_1, \ldots, x_n]$ (this is essentially the Chinese Remainder Theorem); similarly for the mixed-base representation.

The virtue of the modular representation is that the operations of multiplication and division (when the quotient is known to be integral) are very quick (e.g. multiplication takes O(n) operations for integers requiring n moduli p_i) but the disadvantage is that it is difficult to know the sign or size of a number x, or to do division with remainder. The reverse is true for the mixed-base representation; for instance multiplication now takes $O(n^2)$ operations using the obvious algorithm ("fast" multiplication algorithms involve too much bookkeeping to be justified for "reasonable" integers - say those less than 10^{200}).

This suggests that conversion algorithms are important. The basic idea underlying this appendix is that one should perform as much arithmetic as possible by working modulo primes (i.e., using the modular representation for integers). Then one can convert to the mixed-base representation to check on the size of the resulting integers; i.e. to check that the arithmetic has been done modulo sufficiently many primes.

In the following description of how to pass back and forth between the modular and mixed-base representation let rem(x,k,q) be

$$(x_1+x_2P_1+ \ . \ . \ +x_kP_{k-1}) \bmod q$$

for any integers $q > 1$, $k \geq 1$, and sequence $x = [x_1, \ . \ .]$. This can be computed using only single precision arithmetic via Horner's method:

```
Procedure rem(x,k,q)

        s = x_1 mod q
        for(i = k; i>0 ;i = i-1) s = (x_i+p_i s) mod q
        return s
end;
```

where the notation is

for(initialization;termination condition;looping statment)statement.

In order to conserve storage space it is convenient to perform the conversions "in-place". This is easily done for the mixed-base to modular conversion by working from the top down:

Mixed —> Modular

$$\text{for}(i = n;\ i>0; i = i-1)\ x_i = \text{rem}(x,i,p_i)$$

The reverse conversion requires some auxiliary constants; let $q_i = P_{i-1}^{-1} \bmod p_i$.

Modular —> Mixed:

$$\text{for}(i = 2;\ i<= n\ ; i = i+1)\ x_i = q_i(x_i-\text{rem}(x,i-1,p_i)) \bmod p_i.$$

Since these conversion will be used repeatedly the computation of the q_i can be "preconditioned" and done once and for all at the beginning.

Both of these algorithms require $O(n^2)$ operations; nevertheless they were found to be eminently feasible because a) it is often possible to design algorithms to minimize the conversion done, and b) for "reasonable" numbers (say less than 500 digits) the constants implied by the O notation are small so that these algorithms can be coded very efficiently. There are algorithms that take roughly $O(n \log(n))$ operations to perform these conversions, but they are impractical except for extremely large numbers (see [Aho, Hopcroft, and Ullman]). They require extensive bookkeeping and multiword multiplication routines (in fact, their asymptotic efficiency presupposes the so-called "fast" multiplication algorithms referred to above).

We finish this section by showing how to compute the discriminant of a polynomial having integral coefficients:

f(x) is a polynomial of degree n with integral coefficients

r is a small integer whose meaning is described below; 2 is an acceptable value in virtually all applications x is an array (= string of contiguous words)

```
Procedure discriminant(f)

        i:= 0

        do{

                i = i+1

                x_i = discr(f(x),p_i)

                x_i = q_i  (x_i-rem(x,i-1,p_i))

        }

        until{ either x_j = 0 for i-r ≤ j ≤ i

                or x_j = p_j-1 for i-r ≤ j ≤ i

        }

        output(x)

end;
```

Here the routine discr($f(x),p_i$) computes the discriminant of $f(x)$ modulo p_i, and output(x) takes the mixed-base number x and converts it into a decimal representation suitable for human consumption. The computation of a discriminant modulo a prime is isolated into a procedure which is outlined below.

In order to clarify what's happening in this algorithm we introduce some notation. Let x be the discriminant that we are trying to compute; and let x[m] be the unique integer between $-P_m/2$ and $P_m/2$ that is congruent to x modulo P_m. In the procedure above we are dynamically converting from modular representation to the mixed-base representation. Except for a slight quibble due to the possibility of negative numbers, the array in the procedure above contains x[i]. We terminate when x[i] = x[i+r]; this should be thought of as a very good indication that x = x[i]. If all the numbers that arise are random with respect to the p_i then the probability that x[i] = x[i+r] and x ≠ x[i] is w^{-r}, where w is roughly the word size.

Another way to proceed is to compute a priori bounds on the size of the discriminant and then just compute the discriminant modulo sufficiently many primes to "know" the discriminant exactly by the Chinese Remainder Theorem. This has the advantage of involving no uncertainty, but the (big) disadvantage of requiring (on the average) far too many arithmetical operations.

The above technique is of course not suitable for discriminant computations used in producing Table A_5 since one wants to sieve through a large number of polynomials. In that context it is best to explicitly expand the discriminant as a polynomial in the coefficients of the quintic and then use the usual technique (see [Knuth]) for evaluating a polynomial at successive integral arguments.

The calculation of the discriminant of a polynomial modulo a prime is not relevant to the above ideas; but we briefly indicate an efficient approach. Let n be the degree of f(x) and let R(g,h) denote the resultant of the polynomials g and h. Then

$$\mathrm{Discr}(f(x)) = (-1)^{n(n-1)/2}R(f,f').$$

so that we can certainly compute discriminants if we can compute resultants. Let g be a polynomial of degree n and let h be a polynomial of degree $m < n$. Then if we divide h into g to get a remainder r of degree k then

$$g = hq + r.$$

If b is the leading coefficient of h then

$$R(g,h) = b^{n-k}(-1)^{km}R(h,r)$$

so that this can be used recursively to compute R(g,h). Ultimately one uses $R(g,a) = a^n$ for a constant polynomial a.

Linear Systems of Equations

The above idea of computing a large number "dynamically" using computations modulo primes can be applied to the problem of solving systems of equations with integral coefficients. Actually, the restriction to integral coefficients is ultimately unimportant; any system with rational coefficients can be altered to give one with integral coefficients, and computers only know about rational numbers.

For convenience of exposition assume that we are solving a homogeneous system

1) $$Ax = 0$$

where

a) A is an integral matrix with m rows and n columns

b) dim ker A = 1

c) dim ker $A(p_i) = 1$ (where $A(p_i)$ is the reduction of A mod p_i).

The ideas below can be modified to be applicable even in the absence of any of the above assumptions.

Let x be a solution to 1). We can compute x mod p_i using the usual algorithms for solving linear equations (note that there are no subtleties caused by roundoff error or improper pivoting; a number modulo p_i is either zero or nonzero).

Following the notation above, let $x_i[j]$ be the mixed-base representation of the i-th component of the solution vector x modulo P_j. Let x[j] be the corresponding n-tuple of j-tuples. Just as in the discriminant calculation above we dynamically construct x[j] step by step. At each stage we compute x modulo p_{j+1} and then combine this with x[j] to produce x[j+1]. We expect to know x when x[j] = x[j+r] for some reasonable value of r.

Just as in the discriminant calculation above, we can compute some a priori bounds on the size of x; this involves combining Hadamard's bound on the determinant of a matrix with some simple estimates. The resulting bounds are unfortunately exponentially bad. This means that for a "random" large matrix one would do an excessive amount of arithmetic if one uses these bounds.

In the context above we resorted to relying on the odds. In the case of linear equations we can actually give a very sharp bound on the quantity r.

If B is a vector or matrix let $|B|$ be the supremum of the components of B.

Proposition: Let r _be the least integer such that_

$$P_r > n|A| \quad .$$

Then

$$x[j+r] = x[j] \text{ implies that } x = x[j].$$

Remarks: 1) A variant of this proposition can be found in [Bareiss].

2) In a typical application of this result to the systems in this thesis, the entries of A all fit in one word (i.e. were less than p_1) and n was on the order of 100; thus r = 2 is certainly sufficient.

Proof: By the definitions above, there is an integer x' such that

$$x = x[j+r] + P_{j+r}x'.$$

We are assuming that Ax = 0, and that x[j+r] = x[j] so that

$$0 = Ax[j] + P_{j+r}Ax'.$$

Assume that $Ax' \neq 0$. Since Ax' is certainly an integral vector, $|Ax'| \geq 1$. Therefore

$$P_{j+r} \leq P_{j+r}|Ax'| = |Ax[j]| \leq n|A|P_j.$$

Since the sequence of p_i is an increasing sequence we have $P_{j+r}/P_j > P_r$. Therefore the above inequality contradicts our choice of r and we can only conclude that $Ax' = 0$ so that $x = x[j]$. q.e.d.

These ideas have to be modified if the system of equations is not homogeneous, or if the kernel of A has dimension bigger than one, or if some of the moduli are troublesome (i.e. the rank of A modulo p_i is strictly smaller than the rank of A). None of these modifications are difficult.

One less trivial extension of the above ideas involves the situation in which the entries of A are algebraic integers. Again we can solve the system in the residue fields of the prime ideals of the number field involved. It is simplest to work in prime fields (to avoid the difficulties involved in representing finite fields whose order is not prime). To apply the ideas above it is necessary to be able to solve the system of equations modulo each prime ideal above a given rational prime. If we accept these constraints, then we must find rational primes that split completely in the given number field. The easiest fields for which one can describe the primes that split completely are the cyclotomic fields, so we will assume that A has coefficients in the ring of integers of the field generated by N-th roots of unity.

The linear algebra computations needed in this thesis all involve power series whose coefficients were integers in the field of 20-th roots of unity. More generally, any computations with Artin L-series will involve coefficients in some cyclotomic field.

The procedure for solving the system 1) when A has coefficients in the field of N-th roots of unity can be briefly outlined as follows:

1): Choose large p_i that are congruent to 1 mod N, and solve 1) modulo each of the $\varphi(N)$ primes of degree 1 above p_i in the cyclotomic field.

2) Apply the Chinese Remainder Theorem (!) to use these $\varphi(N)$ solutions to determine a solution x modulo p; write each component of x in the form

$$a_1 + a_2 z + a_3 z^2 + \ldots + a_s z^s$$

$(s = \varphi(N))$.

3) To each of the a_i apply the above ideas concerning the dynamic conversion of modular to mixed-base representations of integers.

We conclude with some brief remarks on the efficiency of these modular techniques.

Ordinarily one thinks that an n x n system of equations takes $O(n^3)$ operations to solve. However in the real world (where the time required to do an arithmetic operation on an integer depends heavily on the size of the integer), one can make some reasonable assumptions about the size of the integers that arise and then estimate that the "straightforward" algorithm takes $O(n^5)$ operations (see [Bareiss]). The modular techniques above require only $O(n^4)$ operations. If the integers involved are big enough to justify "fast" multiplication algorithms, then these asymptotically fast multiplication algorithms also lead to $O(n^4)$ solutions to n n systems of equations. These considerations seem to heavily favor the modular techniques. For a much more thorough discussion of this point (and a slightly different conclusion) see [Bareiss].

Appendix 6: Other Cusps

This appendix contains some generalities on fourier expansions of modular forms at arbitrary cusps, and some specific statements about the fourier expansions of newforms. These latter results are unfinished and are a product of some joint work with Jim Weisinger that may appear elsewhere. We profited from conversations with John Tate and Winnie Li. Unexplained notations follow [Ogg, MF] or [Li].

If we write $x/y \in \mathbf{P}^1(\mathbf{Q})$ as $[x,y]$ then $\Gamma = SL_2(\mathbf{Z})$ acts on $\mathbf{P}^1(\mathbf{Q})$ by

$$\begin{bmatrix} a & b \\ c & d \end{bmatrix}[x,y] = [ax+by, cx+dy].$$

The cusps of $X(N)$ are the orbits of the action of $\Gamma(N) \subset \Gamma$ on $\mathbf{P}^1(\mathbf{Q})$. This set of cusps can be identified with

$$\{[x,y]: x,y \in \mathbf{Z}/N\mathbf{Z},\ (x,y) = 1\}$$

where we identify $[-x,-y]$ with $[x,y]$. The cusp ∞ ($= i\infty$) is $[1,0]$. The cusps of $X_1(N)$ (resp. $X_0(N)$) can be identified with the orbits of $\Gamma_1(N)/\Gamma(N)$ (resp. $\Gamma_0(N)/\Gamma(N)$) acting on the set of cusps of $X(N)$. Any cusp of $X_1(N)$ can be put in the form $[x,y]$ where $(x,N) = 1$, and any cusp of $X_0(N)$ can be put in the form $[x,y]$ where $y = 0$ or $y | N$; we will tacitly assume that any cusps of $X_1(N)$ or $X_0(N)$ that occur here are represented in this form.

If $[x,y]$ is a cusp then we say that $d = (y,N)$ is the <u>divisor</u> of the cusp. If $[x,y]$ is a cusp of $X_0(N)$ then by the above convention its divisor is $d = y$. Put $t = (d, N/d)$. Then the ramification of $[x,y] \in X_0(N)$ over ∞ is $e = N/dt$. The ramification of any cusp of $X_1(N)$ that lies over $[x,y] \in X_0(N)$ is $e = t$. The ramification of any cusp of divisor d in the covering $X(N) \longrightarrow X_1(N)$ is $e = d$. All in all we have a sequence of coverings of Riemann surfaces

$$X(N) \longrightarrow X_1(N) \longrightarrow X_0(N) \longrightarrow X(1)$$

for which the total ramification of any cusp of X(N) over ∞ is N. One reference for the above statements on the ramification degrees is [Ogg, RP, p. 222].

If $p = [x,y]$ is a cusp of X(N) then let γ_p denote an element of Γ with

$$\gamma_p(\infty) = p$$

so that

$$\gamma_p = \begin{bmatrix} x & * \\ y & * \end{bmatrix}, \qquad \det \gamma_p = 1.$$

Any two such elements differ by right multiplication by an element of the stabilizer of ∞ in Γ -

$$\{\pm\begin{bmatrix} 1 & m \\ 0 & 1 \end{bmatrix} : m \in \mathbf{Z}\}.$$

If f is a modular form of weight k for a group G that contains $\Gamma(N)$ and p is a cusp of X(N) then define the fourier expansion (= q-expansion) of f at p to be the fourier expansion of $f|\gamma_p$ at ∞. This q-expansion will be denoted f_p. The modular form $f|\gamma_p$ is a modular form for a subgroup of Γ that is conjugate to G. Since $\Gamma(N)$ is a normal subgroup and each cusp of X(N) has ramification e= N over ∞ we conclude that

$$f_p \in \mathbf{C}[q^{(1/N)}], \qquad q = e^{2\pi i z}, \quad z \in \text{upper half plane}.$$

Define the order of zero of f at p, written $v_p(f)$, to be the exponent of the least nonzero $a_n q^{n/N}$ in the power series f_p. Thus

$$v_p(f) \in (1/N)\mathbf{Z}$$

and $v_p(f)$ is independent of the subgroup on which we choose to consider f a modular form for (e.g. $v_p(f)$ is invariant under "pushing" f up trivially). The definition of f_p depends on the choice of γ_p; if γ_p is multiplied on the right by $\left[\begin{smallmatrix}1 & m\\ 0 & 1\end{smallmatrix}\right]$ then $f_p = \Sigma a_n q^{n/N}$ is changed to $\Sigma e^{2\pi i m/N} a_n q^{n/N}$. Thus $v_p(f)$ is independent of the choice of γ_p. Also we observe that if f is a form of type (k,ϵ,N) and p is a cusp of $X_1(N)$ then $v_p(f)$ depends only on the cusp p' of $X_0(N)$ that lies under p.

With the above notation we can now state some facts that are straightforward consequences of the definitions.

<u>Lemma 1: Let</u> h <u>be a cusp form on</u> $X_0(N)$ <u>and let</u> $p = [x,y]$ <u>be a cusp of</u> $X_0(N)$ <u>with ramification</u> e <u>over</u> ∞. <u>Then</u>

$$v_p(h) \geq 1/e \quad .$$

This lemma just expresses the fact that a cusp form must vanish at the cusps together with the fact that the e-th root of a uniformizing parameter at $\infty \in X(1)$ is a uniformizing parameter at a cusp of $X_0(N)$ of ramification e over ∞. Similarly, modular forms on $X_1(N)$ must have q-expansions at a cusp $p \in X_1(N)$ whose possible nonzero terms are restricted by the ramification of p:

<u>Lemma 2: Let</u> $p = [x,y]$ <u>be a cusp of</u> $X_1(N)$ <u>(so that by the above convention</u> $(x,N) = 1$) <u>and let the divisor of</u> p <u>be</u> $d = (y,N)$. <u>Let</u> f <u>be a modular form on</u> $X_1(N)$ <u>with a q-expansion at</u> p <u>of the form</u>

$$f_p = \Sigma a_n q^{n/N}.$$

<u>Then</u>

$$a_n \neq 0 \text{ only if } n \equiv 0 \bmod d.$$

If f is a modular form of type (k,ε,N) then f is a modular form in the usual sense on $X_1(N)$. If $g = \begin{bmatrix} a & b \\ c & d \end{bmatrix}$ is in $\Gamma_0(N)$ the $f|g = \varepsilon(g)f$. This implies that the q-expansion of f at a cusp of $X_1(N)$ must have its nonzero coefficients lying in a certain congruence class.

Lemma 3: Let $[x,y]$ be a cusp of $X_1(N)$ and let f be a form of type (k,ε,N) with

$$f_p = \Sigma a_n q^{n/N}.$$

Put $e = N/dt$ where $d = (y,N)$ is the divisor of p and $t = (d,N/d)$. The stabilizer in $\Gamma_1(N)$ of the cusp p is an infinite cyclic group generated by a matrix $\gamma = \begin{bmatrix} * & * \\ * & 1+xye \end{bmatrix}$. Let a be the number determined by

$$\varepsilon(1+xye) = e^{2\pi i a/t}, \qquad 0<a<t.$$

Then

$$a_n \neq 0 \text{ only if } n \equiv ad \bmod dt.$$

With the notation of the preceding lemma, call a/t the fractional compulsory zero of a form of type (k, ,N) at the cusp p. Similarly, each elliptic point has an associated fractional compulsory zero; consult [Tate, NFZ] for the precise formula.

Proposition ([Tate, NFZ]): Modular forms of type (k,ε,N) can be naturally realized as sections of a line bundle on $X_0(N)$. The degree of this bundle is

$$(k/12)[\,\bar{\Gamma}:\bar{\Gamma}_0(N)] - \Sigma(\text{fractional compulsory zero at } p)$$

where the sum is over all cusps and elliptic points. The degree of the line bundle of cusp forms is the degree above minus the number of cusps with no compulsory zero.

The last basic fact that we need is a formula for $v_p(f|B_d)$. Here we use the notation of [Atkin-Lehner] so that if f is a modular form on $X_1(N)$ then the "push-up" $f|B_d$ is a modular form on $X_1(Nd)$. The operator B_d corresponds to a scalar times the matrix $\left[\begin{smallmatrix} d & 0 \\ 0 & 1 \end{smallmatrix}\right]$ and so B_d takes the cusp [x,y] of $X_1(dN)$ to the cusp [dx,y] of $X_1(N)$; here we must divide through by (d,y) in order to express this cusp in standard form. The point of the next lemma is that the q-expansion of f B_d at [x,y] is related to the q-expansion of f at [dx,y].

Lemma 4: *If* p = [x,y] *is a cusp of* $X_1(N)$ *and* d N *then there is a unique cusp* p' = [x',y'] *of* $X_1(N/d)$ *with*

$$\gamma_{p'}^{-1}\begin{bmatrix} d & 0 \\ 0 & 1 \end{bmatrix}\gamma_p = \begin{bmatrix} * & * \\ 0 & * \end{bmatrix}.$$

Let s = y/y'. *then*

$$v_p(f|B_d) = (s^2/d)v_{p'}(f).$$

If $\chi:(\mathbf{Z}/M\mathbf{Z})^* \longrightarrow \mathbf{C}^*$ is a Dirichlet character and $a \in \mathbf{Z}$ then define

$$\tau(\chi,a) = \Sigma\chi(m)e^{2\pi iam/M}$$

where the sum is extended over all $m \in \mathbf{Z}/M\mathbf{Z}$ and we define $\chi(m) = 0$ if $(m,M) > 1$. Also put $\tau(\chi) = \tau(\chi,1)$, and let $\varphi(n)$ be the Euler phi-function.

If f is a form on $X_1(N)$ with

$$f_\infty = \Sigma a_n q^n$$

and χ is a Dirichlet character then f_χ is the modular form with

$$(f_\chi)_\infty = \Sigma\, \chi(n)a_n q^n.$$

Let B_d operate on power series in the way that it acts on q-expansions at ∞ :

$$B_d g(x) = g(x^d), \qquad g \in \mathbf{C}[x].$$

Now we can give a formula that, in favorable cases, allows one to compute the q-expansions of newforms at arbitrary cusps.

Theorem: Let f be a normalized newform of type (k,ϵ,N) and let g be that form with

$$f|W_N = cg, \quad g_\infty = \Sigma\, a_n q^n, \quad a_1 = 1$$

(so that f and g are an "automorphic pair"). Let $p = [x,y]$ be a cusp of $X_1(N)$ with $(x,N) = 1$. Then

$$B_N(f_p) = cN^{-k/2}\epsilon(-x)\ \Sigma\Sigma\ A(d,\chi)(g_\chi | W_{(N^2/d)})$$

where the first summation is over all divisors d of N, the second sum is the summation over all $\chi:(\mathbf{Z}/(N/d)\mathbf{Z})^* \longrightarrow \mathbf{C}^*$, and

$$A(d,\chi) = a_d d^{-k/2}\chi^2(x)\tau(\bar{\chi},-xy)/\varphi(N/d).$$

Remarks: Note that g_χ is indeed a modular form (not necessarily new) of level $N(N/d) = N^2/d$. If $k = 1$ so that f corresponds to a 2-dimensional galois representation then the "eigenvalues" of the g_χ under the W operators are essentially the constants in the functional equation for the twists of the galois representation by the characters χ. In general it is difficult to determine the behavior of these constants under twisting by a character whose conductor divides the level N.

The main application of this theorem needed in this thesis is to the Eisenstein series produced by Theorem 7. In general the statement of the above theorem presupposes a definition of "newform" for non-cusp forms; here a non-cusp newform of type $(1, \epsilon, N)$ is just a form $f_{\mathbf{T}}$ for $\mathbf{T}$ as in Theorem 7. Thus

$$\mathbf{T} = \chi_1 \oplus \chi_2, \ \mathrm{cond}(\chi_1) = N_1, \ \mathrm{cond}(\chi_2) = N_2, \ N_1 N_2 = N.$$

Corollary: Let $f = f_{\mathbf{T}}$ *be the modular form associated to* χ_1, χ_2, *and let* $f_p = \Sigma a_n q^{n/N}$. *Then*

$$a_0 = (-1/2\pi)[\chi_1(y/N_2)\chi_2(-x)\tau(\chi_2)L(1\ \chi_1\ \overline{\chi}_2)$$

$$+ \quad \chi_2(y/N_1)\chi_1(-x)\tau(\chi_1)L(1, \overline{\chi}_1\ \chi_2)].$$

where $\chi(u) = 0$ *if* χ *is a Dirichlet character and* $u \notin \mathbf{Z}$. *For positive* n *we have*

$$a_n = (-\tau(\chi_1)\tau(\chi_2)/N)[\Sigma\ \chi_1(r)\ \chi_2(s)\zeta^{(r+s+rsxy)}].$$

where the sum is over all r *and* s *in* $\mathbf{Z}/N\mathbf{Z}$, *and* $\zeta = e^{2\pi i/N}$.

Note that the Eisenstein series f of the corollary can be nonzero at a cusp $[x,y]$ only if

$$(y, N) = N_1 \text{ or } N_2.$$

Thus a product of Eisenstein series of this type is "often" a cusp form, which is useful for generating bases of spaces of cusp forms of weight $k \geq 2$.

Appendix 7

Two of Artin's first examples of nonabelian L-series that might not be holomorphic were the 3 and 4 dimensional representations of A_5. The representation in the corollary to Theorem 9 has a holomorphic L-series, but it factors through a representation not of A_5 but rather a central covering group of A_5. Nonetheless, by using some results of Shimura and Rankin it is possible to show that the non-trivial irreducible representations of the galois group of the underlying A_5 extension of $\mathbf{Q}$ must have holomorphic L-series. The purpose of this appendix is to establish a more general fact: whenever there is an icosahedral representation that, together with all of its "twists", satisfies Artin's conjecture then all of the twists of the non-trivial irreducible representations of the galois group of the underlying A_5 field have holomorphic L-series. The argument essentially follows a letter of Serre.

We start by making a definition that simplifies the statements of the theorems in this appendix.

Definition: An irreducible projective galois representation $T:G_{\mathbf{Q}} \longrightarrow PGL_n(\mathbf{C})$ is said to be <u>entire</u> if $L(s,\mathbf{T})$ is holomorphic for all liftings $\mathbf{T}$ of T.

Thus theorem 9 gives us an icosahedral projective representation that is entire. In fact the conjugate projective representation is also entire (any automorphism of $\mathbf{C}$ that takes $\sqrt{5}$ to $-\sqrt{5}$ interchanges the two projective representations and takes an icosahedral newform associated to a lifting of one of them to an icosahedral newform associated to a lifting of the other).

It is easy to check that the tensor product of two projective representations is well defined; similarly for a symmetric tensor power of a projective representation. Also it makes sense to speak of odd (resp. even) two dimensional projective representations (since twisting an ordinary representation by a character λ changes the determinant by λ^2).

Theorem: Let T, $T':G_{\mathbf{Q}} \longrightarrow PGL_2(\mathbf{C})$ be entire two dimensional odd projective representations.

A: (Rankin) If $T \otimes T'$ is irreducible then it is entire.

B: (Shimura) If the second symmetric power $S^2(T)$ is irreducible then it is entire.

If f and f' are newforms of weight one that correspond to liftings $\mathbf{T}$ and $\mathbf{T}'$ of T and T' then the results in [Rankin] (see also [Ogg, CL]) show that $L(s,\mathbf{T}\otimes\mathbf{T}')$ is entire except possibly for poles at s = 0 and s = 1 (in [Rankin] this L-series occurs as the "convolution" of the L-series attached to f and f'). If ϵ is any Dirichlet character then Shimura proves that $L(s,S^2(\mathbf{T})\otimes\epsilon)$ (which is $D(s,f,\epsilon)$ in the notation of [Shimura]) is entire except possibly for poles at s = 0 and s = 1. But it is known that Artin L-series of irreducible representations do not have poles at 0 or 1 so indeed the above theorem is an immediate consequence of the results of Shimura and Rankin.

If T is an irreducible projective representation of A_5 let $c(T) = \pm 1$ denote the element of the Schur multiplier group $H^2(A_5,\mathbf{C}^*) \cong \{\pm 1\}$ attached to T. The projective representations with c(T) = 1 can be uniquely lifted to linear representations of A_5. All of the other projective representations can be lifted to linear representations of a central covering of A_5 by $\{\pm 1\}$. This group will be denoted G and can be realized as the inverse image of $A_5 \subset PGL_2(\mathbf{C}) \cong PSL_2(\mathbf{C})$ in $SL_2(\mathbf{C})$; G is also isomorphic to $SL_2(\mathbf{F}_5)$. For the convenience of the reader the character table of G is given in the table at the end of this appendix (though all of the results here can be obtained without explicitly using this table).

Theorem: Let $T_2:G_{\mathbf{Q}} \longrightarrow PGL_2(\mathbf{C})$ be an odd entire icosahedral projective representation, and let K be the corresponding A_5 extension of $\mathbf{Q}$. Then any non-trivial irreducible projective representation $T:G(K/\mathbf{Q}) \longrightarrow PGL_n(\mathbf{C})$ with c(T) = 1 is entire.

Proof: The five dimensional representation of A_5 is monomial so it suffices to check the three and four dimensional representations.

By the remarks above, the conjugate representation T_2' is also entire. The representation $T = T_2 \otimes T_2'$ has $c(T) = 1$. If T is irreducible it must correspond to the four dimensional representation of A_5. Moreover, by case A of the theorem above, T would then be entire.

The reducibility/irreducibility of T is purely a question about the representations of the finite group G. It can be checked from the character table that if $\mathbf{T}_2$ and $\mathbf{T}_2'$ are the two irreducible two dimensional representations of G then $\mathbf{T}_4 = \mathbf{T}_2 \otimes \mathbf{T}_2'$ is the irreducible four dimensional representation of G with $c(\mathbf{T}_4) = 1$. Alternatively we can observe that a four dimensional representation that factors through a representation of A_5 can be reducible only if it involves the trivial representation. But

$$\mathrm{Hom}_G(1, \mathbf{T}_2 \otimes \mathbf{T}_2') \cong \mathrm{Hom}(\mathbf{T}_2, \mathbf{T}_2') \cong \{0\}$$

(since $\mathbf{T}_2$ is self-dual and $\mathbf{T}_2$ and $\mathbf{T}_2'$ are non-isomorphic irreducible representations) so that indeed T is irreducible. This finishes the proof of the theorem for four dimensional representations.

If $S^2(T)$ (resp. $S^2(T')$) is irreducible then by B of the theorem above it follows that $S^2(T)$ (resp. $S^2(T')$) is entire. Again the irreducibility of $S^2(T)$ can be checked directly by using the character table of G. Alternatively we can observe that $S^2(\mathbf{T}_2)$ is reducible only if it involves the trivial representation. But $\mathbf{T}_2 \otimes \mathbf{T}_2 = \Lambda^2 \mathbf{T}_2 \oplus S^2(\mathbf{T}_2) \cong 1 \oplus S^2(\mathbf{T}_2)$ (since the determinant of $\mathbf{T}_2$ is trivial) and

$$\mathrm{Hom}_G(1, \mathbf{T}_2 \otimes \mathbf{T}_2) \cong \mathrm{Hom}_G(\mathbf{T}_2, \mathbf{T}_2)$$

is one dimensional so that $S^2(\mathbf{T}_2)$ is irreducible. This finishes the proof of the theorem.

Table App7.1

Let G be the inverse image of $A_5 \subset PSL_2(\mathbb{C})$ in $SL_2(\mathbb{C})$ (so that G is isomorphic to $SL_2(F_5)$); G is a central extension of A_5 by the group $\{\pm 1\}$ of order 2. The character table of G is given below; this was one of the first character tables ever computed (Frobenius, Ges. Abh. III, p. 128).

For n = 1,2,3,4,6 let C_n be the unique conjugacy class of G consisting of the elements of order n. Let C_5 and C_5' be the two conjugacy classes of elements of order 5, and if $g \in C_5$ (resp. C_5') let C_{10} (resp. C_{10}') be the conjugacy class of -g. The degree of an irreducible character is indicated by its subscript. If θ is such a character then (in the notation of the text) $c(t) = \theta(-1)/\theta(1)$ so that the representations that factor through representations of A_5 are precisely those whose value on C_2 is positive. The numbers u and v are the distinct roots of $x^2-x-1 = 0$.

Character Table of G :

	C_1	C_2	C_4	C_3	C_6	C_5	C_{10}	C_5'	C_{10}'
θ_1	1	1	1	1	1	1	1	1	1
θ_3	3	3	-1	0	0	u	u	v	v
θ_3'	3	3	-1	0	0	v	v	u	u
θ_4	4	4	0	1	1	-1	-1	-1	-1
θ_5	5	5	1	-1	-1	0	0	0	0
θ_2	2	-2	0	-1	-1	u-1	1-u	v-1	1-v
θ_2'	2	-2	0	-1	-1	v-1	1-v	u-1	1-u
θ_4	4	-4	0	1	-1	-1	1	-1	1
θ_6	6	-6	0	0	0	1	-1	1	-1

Conductors of Quintic Fields

$800 = 2^5 5^2$	[0, 10,-10, 35,-18]	(7, 3; 55000: 5000)	2[17], 5[7], 11[un.]
$837 = 3^3 31^1$	[4, 25, 20, 8, 5]	(2, 5; 77841: 8649)	3[11], 31[1]
$992 = 2^5 31^1$	[9, 20, 36,-11, 1]	(5 7; 46128: 7688)	2[17], 3[un.], 31[1]
$1161 = 3^3 43^1$	[4, 25, 17, 5, 2]	(5, 2; 31347: 387)	3[11], 43[2]
$1188 = 2^2 3^3 11^1$	[4, 13, -5, -2, 1]	(5, 7; 4356: 2178)	2[5], 3[11], 11[1]
$1376 = 2^5 43^1$	[2, 6, 8, 10, 8]	(3, 7; 688: 344)	2[17], 43[2]
$1501 = 19^1 79^1$	[6, 19, 25, 11, 2]	(3, 2; 1501: 1501)	19[3], 79[2]
$1600 = 2^6 5^2$	[5, 20, 0, 15, -1]	(3, 7; 80000: 5000)	2[18], 5[7]
$1687 = 7^1 241^1$	[2, 3, 2, 12,-16]	(3,17; 26992: 1687)	2[un.], 7[3], 241[2]
$1825 = 5^2 73^1$	[4, -1,-21, -1, -7]	(2, 7; 45625: 365)	5[6], 73[2]
$1948 = 2^2 487^1$	[0, 20,-20,-16,-16]	(3, 7;498688: 974)	2[5], 487[3]
$2083 = 2083^1$	[1, 5,-11, 4, 1]	(2,11; 2083: 2083)	2083[3]
$2264 = 2^3 283^1$	[6, 1, 4,-24, 32]	(3,19;289792: 2264)	2[15], 283[2]
$2316 = 2^2 3^1 193^1$	[9, 21, 35, 30, 12]	(5,13; 13896: 1153)	2[5], 3[3], 193[2]
$2336 = 2^5 73^1$	[0, 2, 4, -2, -4]	(3, 7; 1168: 584)	2[17], 73[3]
$2416 = 2^4 151^1$	[0, -2, 2, 5, 6]	(7, 3; 2416: 604)	2[16], 151[2]
$2464 = 2^5 7^1 11^1$	[8, 8,-40, 12, 32]	(13, 3;216832: 6776)	2[17], 7[2], 11[1]
$2577 = 3^1 859^1$	[3, 7, 6,-11,-24]	(5, 2; 23193: 2577)	3[3], 859[2]
$2624 = 2^6 41^1$	[10, -1, -2, 39, 32]	(3,11;255512:13448)	2[18], 19[un.], 41[1]
$2653 = 7^1 379^1$	[3, 1, -4, 17, -8]	(3, 2; 18571: 2653)	7[3], 379[2]
$2673 = 3^5 11^1$	[2, 28,-13, 4, -1]	(2, 5; 29403: 3267)	3[13], 11[1]
$2673 = 3^5 11^1$	[3, -3, -4, 12,-21]	(5, 7; 13068: 3267)	2[un.], 3[13], 11[1]
$2700 = 2^2 3^3 5^2$	[5, 25, 0, 0,-30]	(11,13;798750:11250)	2[5],3[11],5[7],71[un.]
$2707 = 2707^1$	[8, 26, 36, 29,-36]	(5, 3;173248: 2707)	2[un.], 2707[3]
$2721 = 3^1 907^1$	[2, 37, -7, 25, -4]	(5, 2;661203: 2721)	3[3], 907[2]
$2761 = 11^1 251^1$	[4, 13, 17, -2,-32]	(3,17; 60742:30371)	2[un.], 11[1], 251[3]
$2764 = 2^2 691^1$	[5, -8, 14,-16, 8]	(3, 5; 22112: 2764)	2[14], 691[2]
$2863 = 7^1 409^1$	[0, 12,-21, 22, -7]	(2,13; 25767: 2863)	3[un.], 7[3], 409[3]
$2884 = 2^2 7^1 103^1$	[7, 13,-11,-10, 12]	(5, 3; 40376: 1442)	2[5], 7[3], 103[2]
$2884 = 2^2 7^1 103^1$	[1, -1, -1,-12, 16]	(3,23; 11536: 1442)	2[5], 7[3], 103[2]

2979 = $3^{2}331^{1}$	[8, 15,-17, 29, 3]	(2, 7;217467: 2979)	3[10], 73[un.], 331[3]
2981 = $11^{1}271^{1}$	[4, 26, -1, 12, 17]	(2,13;587257: 2981)	11[3], 197[un.], 271[2]
3004 = $2^{2}751^{1}$	[7, 17, 22, 18, 10]	(7, 3; 1502: 1502)	2[5], 751[3]
3031 = $7^{1}433^{1}$	[2, 5, 27, -7, 19]	(2,11;148519: 3031)	7[3], 433[2]
3139 = $43^{1}73^{1}$	[4, 13, 11, 3,-34]	(5, 2; 9417: 3139)	3[un.], 43[3], 73[2]
3168 = $2^{5}3^{2}11^{1}$	[8, 30, 4,-10,-12]	(17, 7;331056: 8712)	2[17],3[10],11[1],19[un.]
3168 = $2^{5}3^{2}11^{1}$	[1, -4, 8, 25, 1]	(5, 7; 5808: 2904)	2[17], 3[6], 11[1]
3184 = $2^{4}199^{1}$	[5, 8,-20,-21, -5]	(3, 7; 12736: 796)	2[16], 199[2]
3203 = 3203^{1}	[1, -1, 9, 20, 11]	(2,17; 3203: 3203)	3203[3]
3348 = $2^{2}3^{3}31^{1}$	[3,-15, 22,-12, 6]	(7, 5; 17298:17298)	2[5], 3[11], 31[1]
3377 = $11^{1}307^{1}$	[1, -5, 14,-11, 4]	(3, 2; 3377: 3377)	11[3], 307[2]
3396 = $2^{2}3^{1}283^{1}$	[9, 9, 23, 18, 12]	(5,11; 81504: 1698)	2[5], 3[3], 283[2]
3423 = $3^{1}7^{1}163^{1}$	[10, 3, -2,-14, 9]	(2,17; 92421: 3423)	3[3], 7[2], 163[2]
3431 = $47^{1}73^{1}$	[9, 19, 7,-18, 29]	(2,29;250463: 3431)	47[3], 73[20]
3547 = 3547^{1}	[10, 32, 34, 23,-36]	(5, 3;227008: 3547)	2[un.], 3547[3]
3548 = $2^{2}887^{1}$	[2, 17, 4,-24,-16]	(3, 7;113536: 1774)	2[5], 887[3]
3556 = $2^{2}7^{1}127^{1}$	[3, 9, -6, -4,-40]	(3, 5; 99568: 1778)	2[5], 7[3], 127[2]
3587 = $17^{1}211^{1}$	[3, -1,-24, 13, 17]	(2,13; 32283: 3587)	3[un.], 17[3], 211[3]
3627 = $3^{2}13^{1}31^{1}$	[8, 27, 31, 17, 3]	(2,19; 3627: 3627)	3[10], 13[2], 31[3]
3676 = $2^{2}919^{1}$	[5, 2, 14, -3, 29]	(5, 3;117632: 1838)	2[5], 919[3]
3708 = $2^{2}3^{2}103^{1}$	[5, 3,-22, -8, -6]	(5,11; 50058: 1854)	2[5], 3[10], 103[3]
3756 = $2^{2}3^{1}313^{1}$	[0, -3, 10, 30,-18]	(7, 5;101412: 1878)	2[5], 3[3], 313[2]
3775 = $5^{2}151^{1}$	[0, -3, -1, 1, -3]	(2, 3; 755: 755)	5[5], 151[3]
3861 = $3^{3}11^{1}13^{1}$	[4, 13, 39, 9, 12]	(5, 2;184041:14157)	3[11], 11[1], 13[20]
3875 = $5^{3}31^{1}$	[0, 5, 10, 0, 1]	(2, 7; 3875: 3875)	5[8], 31[3]
3981 = $3^{1}1327^{1}$	[3, -3, -7, 18, 15]	(2, 5; 11943: 3981)	3[3], 1327[2]
4000 = $2^{5}5^{3}$	[5, 20,-20, 0, 8]	(3, 7;200000: 1000)	2[17], 5[8]
4027 = 4027^{1}	[1, 9,-38, 13, 23]	(2, 7; 12081: 4027)	3[un.], 4027[3]
4108 = $2^{2}13^{1}79^{1}$	[4, 3, 9, 4, 5]	(3, 5; 2054: 2054)	2[5], 13[2], 79[3]
4156 = $2^{2}1039^{1}$	[1, 33,-27,-36, -8]	(5,13; 49872: 4156)	2[14], 3[un.], 1039[2]
4161 = $3^{1}19^{1}73^{1}$	[8, 14, 13, 12, 9]	(2, 7; 12483: 4161)	3[3], 19[2], 73[2]

4207 = $7^{1}601^{1}$ [5, 9, 17,-16, 33] (2,11;206143: 4207) 7[3], 601[2]

4237 = $19^{1}223^{1}$ [7,-15, 22, -9, 4] (3, 2; 29659: 4237) 7[un.], 19[3], 223[2]

4256 = $2^{5}7^{1}19^{1}$ [0, 6,-12, 5, -4] (3,11; 8512: 1064) 2[17], 7[2], 19[2]

4269 = $3^{1}1423^{1}$ [4, 4,-23, 34,-23] (2,31;115263: 4269) 3[3], 1423[2]

4288 = $2^{6}67^{1}$ [4, 5, 8, 3, 2] (3, 5; 536: 536) 2[18], 67[2]

4328 = $2^{3}541^{1}$ [1, 23, 31, 10, -2] (3,19; 60592: 4328) 2[15], 7[un.], 541[2]

4400 = $2^{4}5^{2}11^{1}$ [11, 33, 33, 11, 11] (3, 7; 12100: 2420) 2[16], 5[5], 11[1]

4428 = $2^{2}3^{3}41^{1}$ [4, 31, 33,-36, 9] (11, 7;453870:30258)2[5],3[11],5[un.],41[1]

4588 = $2^{2}31^{1}37^{1}$ [5, 7, 3, 4,-16] (3, 5; 18352: 2294) 2[5], 31[3], 37[2]

4763 = $11^{1}433^{1}$ [1, 1, 10, 11, 33] (2, 3; 61919: 4763) 11[3], 13[un.], 433[2]

4836 = $2^{2}3^{1}13^{1}31^{1}$ [1, 19,-21,-30, 36] (5,11;449748:74958) 2[5],3[3],13[2],31[1]

4904 = $2^{3}613^{1}$ [9,-14, 2, 16, 16] (7, 3;510016: 4904) 2[15], 13[un.], 613[2]

4911 = $3^{1}1637^{1}$ [10, 30, 39, 18, 27] (2, 5;132597: 4911) 3[3], 1637[3]

4975 = $5^{2}199^{1}$ [6, 17, 34, 38, -1] (2,17; 24875: 995) 5[6], 199[2]

5024 = $2^{5}157^{1}$ [6, 14, 20, 26, 16] (3,47; 2512: 1256) 2[17], 157[2]

5031 = $3^{2}13^{1}43^{1}$ [3, 2, 5, 27, 25] (2, 5; 15093: 5031) 3[10], 13[2], 43[3]

5044 = $2^{2}13^{1}97^{1}$ [4, 19,-14, 14, 40] (3, 5;1049152: 5044) 2[14], 13[20], 97[3]

5171 = 5171^{1} [5, 8, 33, 8, 28] (3,11;144788: 5171) 2[un.], 7[un.], 5171[3]

5203 = $11^{2}43^{1}$ [0, -3, 7, -5, 3] (2, 3; 473: 473) 11[5], 43[3]

5225 = $5^{2}11^{1}19^{1}$ [10, 29, 14, -8,-23] (2, 7;103455:11495) 3[un.],5[6],11[1],19[2]

5239 = $13^{2}31^{1}$ [2, 12, 9, 28,-11] (2,41; 403: 403) 13[6], 31[2]

5239 = $13^{2}31^{1}$ [0, 0,-31, 31, 31] (3,11; 24986:12493) 2[un.], 13[6], 31[1]

5287 = $17^{1}311^{1}$ [5, 32, 9, 22, 24] (5, 3;1089122:5287)2[un.],17[3],103[un.],311[3]

5332 = $2^{2}31^{1}43^{1}$ [5, 3, 30, -8, 12] (5, 3;165292: 2666) 2[5], 31[2], 43[3]

5357 = $11^{1}487^{1}$ [7, 14,-21, -9, -1] (2, 7; 16071: 5357) 3[un.], 11[3], 487[2]

5369 = $7^{1}13^{1}59^{1}$ [3, 3, 0, -7, 7] (2,83; 5369: 5369) 7[2], 13[2], 59[3]

5373 = $3^{3}199^{1}$ [3, 6, -5, -9,-12] (5,13; 14328: 1791) 2[un.], 3[11], 199[2]

5373 = $3^{3}199^{1}$ [2, 1, 7, 23,-11] (2,11; 48357: 1791) 3[11], 199[2]

5373 = $3^{3}199^{1}$ [3, 0, -5, 15,-18] (5,13; 32238: 1791) 2[un.], 3[11], 199[2]

5373 = $3^{3}199^{1}$ [1, -2, 3, 9, 9] (2,11; 5373: 1791) 3[11], 199[2]

5373 = $3^{3}199^{1}$ [2, -2, -3, 18, 36] (5,13; 32238: 1791) 2[un.], 3[11], 199[2]

$5373 = 3^{3}199^{1}$	[8, 25, 31, 23, 37]	(2,11; 48357: 1791)	3[11], 199[2]
$5373 = 3^{3}199^{1}$	[2, 4, 15, 12, 12]	(5,13; 14328: 1791)	2[un.], 3[11], 199[2]
$5425 = 5^{2}7^{1}31^{1}$	[1, -8, 13,-25, 25]	(2,11; 27125: 1085)	5[5], 7[3], 31[2]
$5603 = 13^{1}431^{1}$	[2, -3, 0, 6,-17]	(2, 5; 5603: 5603)	13[2], 431[3]
$5619 = 3^{1}1873^{1}$	[2, 38,-35,-18, 27]	(2,37;2039697: 5619)	3[3], 11[un.], 1873[2]
$5697 = 3^{3}211^{1}$	[1, 19, 25, 10, 1]	(2,19; 1899: 1899)	3[11], 211[2]
$5789 = 7^{1}827^{1}$	[8, 18,-21, -6, 7]	(2,19; 40523: 5789)	7[2], 827[3]
$5984 = 2^{5}11^{1}17^{1}$	[15,-20,-16, 20, 12]	(7, 3;131648:16456)	2[17], 11[1], 17[3]
$6021 = 3^{3}223^{1}$	[0, 3, 15, -9, 36]	(5, 2; 18063: 2007)	3[11], 223[2]
$6123 = 3^{1}13^{1}157^{1}$	[11, 2, 31, 39, 15]	(2, 5;569439:6123)	3[3],13[3],31[un.],157[2]
$6163 = 6163^{1}$	[7,-11, 34,-25, 7]	(2,11;117097: 6163)	19[un.], 6163[3]
$6176 = 2^{5}193^{1}$	[5, 0,-16, -4, 20]	(7, 3; 12352: 1544)	2[17], 193[3]
$6179 = 37^{1}167^{1}$	[6, 31, 11, -4, 4]	(3,11;173012:6179)	2[un.],7[un.],37[3],167[3]
$6232 = 2^{3}19^{1}41^{1}$	[4, 9, 10,-10, 4]	(5, 3; 12464: 6232)	2[15], 19[2], 41[3]
$6244 = 2^{2}7^{1}223^{1}$	[2, 19, -8, 18, -4]	(5,23; 99904: 6244)	2[14], 7[2], 223[2]
$6289 = 19^{1}331^{1}$	[3, 32, -1, 15, 11]	(2,11;672923: 6289)	19[2], 107[un.], 331[3]
$6349 = 7^{1}907^{1}$	[5, 21,-40, 11, 8]	(11, 2;311101: 6349)	7[3], 907[2]
$6588 = 2^{2}3^{3}61^{1}$	[9, 12,-38, 27, -3]	(5,11; 26352: 1098)	2[5], 3[11], 61[3]
$6652 = 2^{2}1663^{1}$	[2, 11, 10,-12, 8]	(3, 5; 53216: 3326)	2[5], 1663[3]
$6677 = 11^{1}607^{1}$	[4, 0,-19,-18,-11]	(2,17; 20031: 6677)	3[un.], 11[3], 607[2]
$6683 = 41^{1}163^{1}$	[3,-11, -7, 30, 25]	(2,19; 33415: 6683)	5[un.], 41[3], 163[3]
$6913 = 31^{1}223^{1}$	[7, 2,-39, 31,-18]	(5, 3;110608: 6913)	2[un.], 31[3], 223[2]
$6935 = 5^{1}19^{1}73^{1}$	[9, 4,-25, -5, 25]	(2, 3; 34675: 6935)	5[20], 19[3], 73[2]
$7047 = 3^{5}29^{1}$	[1, 7, 2, 11,-13]	(2,11; 7047: 783)	3[13], 29[3]
$7064 = 2^{3}883^{1}$	[7, 2, 18,-16, 16]	(5, 7;169536: 7064)	2[15], 3[un.], 883[2]
$7075 = 5^{2}283^{1}$	[3, 5, 5, 8, 1]	(2,17; 1415: 1415)	5[6], 283[2]
$7236 = 2^{2}3^{3}67^{1}$	[9, 27, 34, 18, 6]	(7, 5; 1206: 1206)	2[5], 3[11], 67[2]
$7324 = 2^{2}1831^{1}$	[9, 34,-22, 17,-15]	(7, 3;937472: 7324)	2[14], 1831[2]
$7363 = 37^{1}199^{1}$	[1,-13, 39, 32, 12]	(5, 3;471232: 7363)	2[un.], 37[2], 199[3]
$7372 = 2^{2}19^{1}97^{1}$	[3, 5, 13, 36, 18]	(5,13;44232: 7372)	2[14],3[un.],19[2],97[2]
$7409 = 31^{1}239^{1}$	[0, -5, 31,-31, 31]	(2,13;111135:7409)	3[un.],5[un.],31[2],239[3]

7479 = $3^{3}277^{1}$ [4, 4, 3, 0, 9] (2, 7; 7479: 2493) 3[11], 277[2]

7533 = $3^{5}31^{1}$ [1, 7, 10, 13, 22] (5, 2; 2511: 837) 3[13], 31[2]

7600 = $2^{4}5^{2}19^{1}$ [8, 11,-20, 9, -2] (3, 7; 1520: 380) 2[16], 5[5], 19[2]

7683 = $3^{1}13^{1}197^{1}$[3, 4, -9, 16,-12] (5,23;30732: 7683) 2[un.],3[3],13[2],197[3]

7808 = $2^{7}61^{1}$ [0, 0, 4, 1, -4] (3,11; 976: 976) 2[19], 61[2]

7839 = $3^{2}13^{1}67^{1}$ [7, 21, 22, -5, 6] (5, 2; 23517: 2613) 3[6], 13[2], 67[3]

7915 = $5^{1}1583^{1}$ [6, 13,-11,-13, 33] (2, 3;197875: 7915) 5[20], 1583[3]

7924 = $2^{2}7^{1}283^{1}$ [2, 9, 3, 14,-31] (3, 5; 27734: 3962) 2[5], 7[3], 283[2]

**7947 = $3^{2}883^{1}$ [5, -7,-11, 10, 3] (2, 7; 2649: 2649) 3[6], 883[2]

7999 = $19^{1}421^{1}$ [1, 29,-35, 4, 9] (2,43;407949:7999)3[un.],17[un.],19[3],421[2]

8037 = $3^{2}19^{1}47^{1}$ [3, 33, -8, 21, 9] (2,17;795663:8037) 3[10],11[un.],19[2],47[3]

8103 = $3^{1}37^{1}73^{1}$ [1, 11, -1, 18, 3] (2,19; 24309: 8103) 3[3], 37[2], 73[2]

8135 = $5^{1}1627^{1}$ [7, 11, 15, 30, 25] (2,11; 40675: 8135) 5[20], 1627[3]

8156 = $2^{2}2039^{1}$ [4,-12, 28, 32, 16] (5, 3;521984: 4078) 2[5], 2039[3]

**8164 = $2^{2}13^{1}157^{1}$[8,-23,-29, -2, 1] (3,11;106132: 4082) 2[5], 13[20], 157[2]

8193 = $3^{1}2731^{1}$ [3, -8, 33, 1, 6] (5,13;196632: 8193) 2[un.], 3[3], 2731[2]

8235 = $3^{3}5^{1}61^{1}$ [6, 12, 13, 15,-15] (7,37;27450:2745) 2[un.],3[11],5[20],61[2]

8316 = $2^{2}3^{3}7^{1}11^{1}$[2, -5, 9,-18, 21] (13, 5; 45738:15246) 2[5],3[11],7[2],11[1]

8375 = $5^{3}67^{1}$ [5, 15, 20, 25, 8] (3, 2; 8375: 8375) 5[8], 67[3]

8491 = $7^{1}1213^{1}$ [3, 23,-13,-32, 37] (2, 5;1027411: 8491) 7[3], 11[un.], 1213[2]

8657 = $11^{1}787^{1}$ [1, 15, 39, 36, 11] (2,29; 8657: 8657) 11[3], 787[2]

**8705 = $5^{1}1741^{1}$ [11,-19,-15, 16, -3] (2, 3; 43525: 8705) 5[20], 1741[2]

8708 = $2^{2}7^{1}311^{1}$ [8, 3, 32,-10, 2] (5, 3;121912: 4354) 2[5], 7[2], 311[3]

8756 = $2^{2}11^{1}199^{1}$[7, 24, -6, 31, -1] (3, 7;385264:48158) 2[5], 11[1], 199[3]

8775 = $3^{3}5^{2}13^{1}$ [4,-11, 35,-31, 29] (2,23; 47385: 585) 3[11], 5[5], 13[2]

8800 = $2^{5}5^{2}11^{1}$ [7, 24, 16, 20,-12] (13, 7;193600: 4840) 2[17], 5[5], 11[1]

8879 = $13^{1}683^{1}$ [2, 15, 26, 26, 39] (2,19;150943: 8879) 13[2], 17[un.], 683[3]

9032 = $2^{3}1129^{1}$ [5,-28, 32, 4, 12] (5,17;433536: 9032) 2[15], 3[un.], 1129[2]

9076 = $2^{2}2269^{1}$ [2, 14, -6, -3, 8] (3, 7; 72608: 9076) 2[14], 2269[3]

9099 = $3^{3}337^{1}$ [5, -5, 21,-15, 3] (7, 5; 12132: 3033) 2[un.], 3[11], 337[2]

9121 = $7^{1}1303^{1}$ [0, 23, 15, -5,-21] (2, 5;574623: 9121) 3[un.], 7[2], 1303[3]

9188	$= 2^2 2297^1$	[7, 27,-25,-12, 18]	(5,11;661536: 9188)	2[14], 3[un.], 2297[3]
9219	$= 3^1 7^1 439^1$	[0, -8,-33, 40,-39]	(2,13;728301:9219)	3[3],7[2],79[un.],439[2]
9248	$= 2^5 17^2$	[7, 22, 34, 17,-17]	(3,11; 8704: 136)	2[17], 17[5]
9268	$= 2^2 7^1 331^1$	[0,-11, 7, 14, -7]	(3,13; 9268: 4634)	2[5], 7[2], 331[2]
9500	$= 2^2 5^3 19^1$	[5, 5,-10,-10, 18]	(7, 3; 4750: 4750)	2[5], 5[8], 19[3]
9537	$= 3^1 11^1 17^2$	[10, 29,-19, 3, 21]	(2, 7;771375:6171)	3[3],5[un.],11[1],17[5]
9589	$= 43^1 223^1$	[1, -4, 1, 31,-39]	(2,13; 86301: 9589)	3[un.], 43[3], 223[2]
9632	$= 2^5 7^1 43^1$	[6, 24, 8, 28,-24]	(5, 3;539392: 2408)	2[17], 7[2], 43[2]
9747	$= 3^3 19^2$	[1,-11, 32,-31, 17]	(2,13; 29241: 3249)	3[11], 19[4]
9747	$= 3^3 19^2$	[9, 21, -4,-15, 6]	(5, 2; 9747: 3249)	3[11], 19[4]
9747	$= 3^3 19^2$	[2, 13,-10, -2, -7]	(2,13; 68229: 3249)	3[11], 7[un.], 19[4]
9747	$= 3^3 19^2$	[2, 13, 9, 36, 12]	(5,11; 38988: 3249)	2[un.], 3[11], 19[4]
9747	$= 3^3 19^2$	[5, 10, -9,-33, 39]	(2,13;133209: 3249)	3[11], 19[4], 41[un.]
9760	$= 2^5 5^1 61^1$	[9, 0, 20, 25, 25]	(3, 7;610000: 2440)	2[17], 5[20], 61[2]
9777	$= 3^1 3259^1$	[3, 2, 37,-15, 11]	(2, 7;263979: 9777)	3[3], 3259[2]
9856	$= 2^7 7^1 11^1$	[8, 8,-40, 1, 32]	(3, 5; 94864:13552)	2[19], 7[20], 11[1]
9900	$= 2^2 3^2 5^2 11^1$	[5, 21, -1, 16, 12]	(7,17;261360:10890)	2[5],3[10],5[6],11[1]
9907	$= 9907^1$	[6, 23,-13, 25, -8]	(3, 2;287303: 9907)	29[un.], 9907[3]
9932	$= 2^2 13^1 191^1$	[10, 25, -2, 20, -6]	(5, 3; 79456: 4966)	2[5], 13[2], 191[3]
9959	$= 23^1 433^1$	[11, 37, 11,-12, 9]	(2, 5;209139:9959)	3[un.],7[un.],23[3],433[2]

INDEX

Vol. 489: J. Bair and R. Fourneau, Etude Géométrique des Espaces Vectoriels. Une Introduction. VII, 185 pages. 1975.

Vol. 490: The Geometry of Metric and Linear Spaces. Proceedings 1974. Edited by L. M. Kelly. X, 244 pages. 1975.

Vol. 491: K. A. Broughan, Invariants for Real-Generated Uniform Topological and Algebraic Categories. X, 197 pages. 1975.

Vol. 492: Infinitary Logic: In Memoriam Carol Karp. Edited by D. W. Kueker. VI, 206 pages. 1975.

Vol. 493: F. W. Kamber and P. Tondeur, Foliated Bundles and Characteristic Classes. XIII, 208 pages. 1975.

Vol. 494: A Cornea and G. Licea. Order and Potential Resolvent Families of Kernels. IV, 154 pages. 1975.

Vol. 495: A. Kerber, Representations of Permutation Groups II. V, 175 pages. 1975.

Vol. 496: L. H. Hodgkin and V. P. Snaith, Topics in K-Theory. Two Independent Contributions. III, 294 pages. 1975.

Vol. 497: Analyse Harmonique sur les Groupes de Lie. Proceedings 1973–75. Edité par P. Eymard et al. VI, 710 pages. 1975.

Vol. 498: Model Theory and Algebra. A Memorial Tribute to Abraham Robinson. Edited by D. H. Saracino and V. B. Weispfenning. X, 463 pages. 1975.

Vol. 499: Logic Conference, Kiel 1974. Proceedings. Edited by G. H. Müller, A. Oberschelp, and K. Potthoff. V, 651 pages 1975.

Vol. 500: Proof Theory Symposion, Kiel 1974. Proceedings. Edited by J. Diller and G. H. Müller. VIII, 383 pages. 1975.

Vol. 501: Spline Functions, Karlsruhe 1975. Proceedings. Edited by K. Böhmer, G. Meinardus, and W. Schempp. VI, 421 pages. 1976.

Vol. 502: János Galambos, Representations of Real Numbers by Infinite Series. VI, 146 pages. 1976.

Vol. 503: Applications of Methods of Functional Analysis to Problems in Mechanics. Proceedings 1975. Edited by P. Germain and B. Nayroles. XIX, 531 pages. 1976.

Vol. 504: S. Lang and H. F. Trotter, Frobenius Distributions in GL_2-Extensions. III, 274 pages. 1976.

Vol. 505: Advances in Complex Function Theory. Proceedings 1973/74. Edited by W. E. Kirwan and L. Zalcman. VIII, 203 pages. 1976.

Vol. 506: Numerical Analysis, Dundee 1975. Proceedings. Edited by G. A. Watson. X, 201 pages. 1976.

Vol. 507: M. C. Reed, Abstract Non-Linear Wave Equations. VI, 128 pages. 1976.

Vol. 508: E. Seneta, Regularly Varying Functions. V, 112 pages. 1976.

Vol. 509: D. E. Blair, Contact Manifolds in Riemannian Geometry. VI, 146 pages. 1976.

Vol. 510: V. Poènaru, Singularités C^∞ en Présence de Symétrie. V, 174 pages. 1976.

Vol. 511: Séminaire de Probabilités X. Proceedings 1974/75. Edité par P. A. Meyer. VI, 593 pages. 1976.

Vol. 512: Spaces of Analytic Functions, Kristiansand, Norway 1975. Proceedings. Edited by O. B. Bekken, B. K. Øksendal, and A. Stray. VIII, 204 pages. 1976.

Vol. 513: R. B. Warfield, Jr. Nilpotent Groups. VIII, 115 pages. 1976.

Vol. 514: Séminaire Bourbaki vol. 1974/75. Exposés 453 – 470. IV, 276 pages. 1976.

Vol. 515: Bäcklund Transformations. Nashville, Tennessee 1974. Proceedings. Edited by R. M. Miura. VIII, 295 pages. 1976.

Vol. 516: M. L. Silverstein, Boundary Theory for Symmetric Markov Processes. XVI, 314 pages. 1976.

Vol. 517: S. Glasner, Proximal Flows. VIII, 153 pages. 1976.

Vol. 518: Séminaire de Théorie du Potentiel, Proceedings Paris 1972–1974. Edité par F. Hirsch et G. Mokobodzki. VI, 275 pages. 1976.

Vol. 519: J. Schmets, Espaces de Fonctions Continues. XII, 150 pages. 1976.

Vol. 520: R. H. Farrell, Techniques of Multivariate Calculation. X, 337 pages. 1976.

Vol. 521: G. Cherlin, Model Theoretic Algebra – Selected Topics. IV, 234 pages. 1976.

Vol. 522: C. O. Bloom and N. D. Kazarinoff, Short Wave Radiation Problems in Inhomogeneous Media: Asymptotic Solutions. V. 104 pages. 1976.

Vol. 523: S. A. Albeverio and R. J. Høegh-Krohn, Mathematical Theory of Feynman Path Integrals. IV, 139 pages. 1976.

Vol. 524: Séminaire Pierre Lelong (Analyse) Année 1974/75. Edité par P. Lelong. V, 222 pages. 1976.

Vol. 525: Structural Stability, the Theory of Catastrophes, and Applications in the Sciences. Proceedings 1975. Edited by P. Hilton. VI, 408 pages. 1976.

Vol. 526: Probability in Banach Spaces. Proceedings 1975. Edited by A. Beck. VI, 290 pages. 1976.

Vol. 527: M. Denker, Ch. Grillenberger, and K. Sigmund, Ergodic Theory on Compact Spaces. IV, 360 pages. 1976.

Vol. 528: J. E. Humphreys, Ordinary and Modular Representations of Chevalley Groups. III, 127 pages. 1976.

Vol. 529: J. Grandell, Doubly Stochastic Poisson Processes. X, 234 pages. 1976.

Vol. 530: S. S. Gelbart, Weil's Representation and the Spectrum of the Metaplectic Group. VII, 140 pages. 1976.

Vol. 531: Y.-C. Wong, The Topology of Uniform Convergence on Order-Bounded Sets. VI, 163 pages. 1976.

Vol. 532: Théorie Ergodique. Proceedings 1973/1974. Edité par J.-P. Conze and M. S. Keane. VIII, 227 pages. 1976.

Vol. 533: F. R. Cohen, T. J. Lada, and J. P. May, The Homology of Iterated Loop Spaces. IX, 490 pages. 1976.

Vol. 534: C. Preston, Random Fields. V, 200 pages. 1976.

Vol. 535: Singularités d'Applications Differentiables. Plans-sur-Bex. 1975. Edité par O. Burlet et F. Ronga. V, 253 pages. 1976.

Vol. 536: W. M. Schmidt, Equations over Finite Fields. An Elementary Approach. IX, 267 pages. 1976.

Vol. 537: Set Theory and Hierarchy Theory. Bierutowice, Poland 1975. A Memorial Tribute to Andrzej Mostowski. Edited by W. Marek, M. Srebrny and A. Zarach. XIII, 345 pages. 1976.

Vol. 538: G. Fischer, Complex Analytic Geometry. VII, 201 pages. 1976.

Vol. 539: A. Badrikian, J. F. C. Kingman et J. Kuelbs, Ecole d'Eté de Probabilités de Saint Flour V-1975. Edité par P.-L. Hennequin. IX, 314 pages. 1976.

Vol. 540: Categorical Topology, Proceedings 1975. Edited by E. Binz and H. Herrlich. XV, 719 pages. 1976.

Vol. 541: Measure Theory, Oberwolfach 1975. Proceedings. Edited by A. Bellow and D. Kölzow. XIV, 430 pages. 1976.

Vol. 542: D. A. Edwards and H. M. Hastings, Čech and Steenrod Homotopy Theories with Applications to Geometric Topology. VII, 296 pages. 1976.

Vol. 543: Nonlinear Operators and the Calculus of Variations, Bruxelles 1975. Edited by J. P. Gossez, E. J. Lami Dozo, J. Mawhin, and L. Waelbroeck, VII, 237 pages. 1976.

Vol. 544: Robert P. Langlands, On the Functional Equations Satisfied by Eisenstein Series. VII, 337 pages. 1976.

Vol. 545: Noncommutative Ring Theory. Kent State 1975. Edited by J. H. Cozzens and F. L. Sandomierski. V, 212 pages. 1976.

Vol. 546: K. Mahler, Lectures on Transcendental Numbers. Edited and Completed by B. Diviš and W. J. Le Veque. XXI, 254 pages. 1976.

Vol. 547: A. Mukherjea and N. A. Tserpes, Measures on Topological Semigroups: Convolution Products and Random Walks. V, 197 pages. 1976.

Vol. 548: D. A. Hejhal, The Selberg Trace Formula for PSL $(2,\mathbb{R})$. Volume I. VI, 516 pages. 1976.

Vol. 549: Brauer Groups, Evanston 1975. Proceedings. Edited by D. Zelinsky. V, 187 pages. 1976.

Vol. 550: Proceedings of the Third Japan – USSR Symposium on Probability Theory. Edited by G. Maruyama and J. V. Prokhorov. VI, 722 pages. 1976.

Vol. 551: Algebraic K-Theory, Evanston 1976. Proceedings. Edited by M. R. Stein. XI, 409 pages. 1976.

Vol. 552: C. G. Gibson, K. Wirthmüller, A. A. du Plessis and E. J. N. Looijenga. Topological Stability of Smooth Mappings. V, 155 pages. 1976.

Vol. 553: M. Petrich, Categories of Algebraic Systems. Vector and Projective Spaces, Semigroups, Rings and Lattices. VIII, 217 pages. 1976.

Vol. 554: J. D. H. Smith, Mal'cev Varieties. VIII, 158 pages. 1976.

Vol. 555: M. Ishida, The Genus Fields of Algebraic Number Fields. VII, 116 pages. 1976.

Vol. 556: Approximation Theory. Bonn 1976. Proceedings. Edited by R. Schaback and K. Scherer. VII, 466 pages. 1976.

Vol. 557: W. Iberkleid and T. Petrie, Smooth S^1 Manifolds. III, 163 pages. 1976.

Vol. 558: B. Weisfeiler, On Construction and Identification of Graphs. XIV, 237 pages. 1976.

Vol. 559: J.-P. Caubet, Le Mouvement Brownien Relativiste. IX, 212 pages. 1976.

Vol. 560: Combinatorial Mathematics, IV, Proceedings 1975. Edited by L. R. A. Casse and W. D. Wallis. VII, 249 pages. 1976.

Vol. 561: Function Theoretic Methods for Partial Differential Equations. Darmstadt 1976. Proceedings. Edited by V. E. Meister, N. Weck and W. L. Wendland. XVIII, 520 pages. 1976.

Vol. 562: R. W. Goodman, Nilpotent Lie Groups: Structure and Applications to Analysis. X, 210 pages. 1976.

Vol. 563: Séminaire de Théorie du Potentiel. Paris, No. 2. Proceedings 1975–1976. Edited by F. Hirsch and G. Mokobodzki. VI, 292 pages. 1976.

Vol. 564: Ordinary and Partial Differential Equations, Dundee 1976. Proceedings. Edited by W. N. Everitt and B. D. Sleeman. XVIII, 551 pages. 1976.

Vol. 565: Turbulence and Navier Stokes Equations. Proceedings 1975. Edited by R. Temam. IX, 194 pages. 1976.

Vol. 566: Empirical Distributions and Processes. Oberwolfach 1976. Proceedings. Edited by P. Gaenssler and P. Révész. VII, 146 pages. 1976.

Vol. 567: Séminaire Bourbaki vol. 1975/76. Exposés 471–488. IV, 303 pages. 1977.

Vol. 568: R. E. Gaines and J. L. Mawhin, Coincidence Degree, and Nonlinear Differential Equations. V, 262 pages. 1977.

Vol. 569: Cohomologie Etale SGA 4½. Séminaire de Géométrie Algébrique du Bois-Marie. Edité par P. Deligne. V, 312 pages. 1977.

Vol. 570: Differential Geometrical Methods in Mathematical Physics, Bonn 1975. Proceedings. Edited by K. Bleuler and A. Reetz. VIII, 576 pages. 1977.

Vol. 571: Constructive Theory of Functions of Several Variables, Oberwolfach 1976. Proceedings. Edited by W. Schempp and K. Zeller. VI, 290 pages. 1977

Vol. 572: Sparse Matrix Techniques, Copenhagen 1976. Edited by V. A. Barker. V, 184 pages. 1977.

Vol. 573: Group Theory, Canberra 1975. Proceedings. Edited by R. A. Bryce, J. Cossey and M. F. Newman. VII, 146 pages. 1977.

Vol. 574: J. Moldestad, Computations in Higher Types. IV, 203 pages. 1977.

Vol. 575: K-Theory and Operator Algebras, Athens, Georgia 1975. Edited by B. B. Morrel and I. M. Singer. VI, 191 pages. 1977.

Vol. 576: V. S. Varadarajan, Harmonic Analysis on Real Reductive Groups. VI, 521 pages. 1977.

Vol. 577: J. P. May, E_∞ Ring Spaces and E_∞ Ring Spectra. IV, 268 pages. 1977.

Vol. 578: Séminaire Pierre Lelong (Analyse) Année 1975/76. Edité par P. Lelong. VI, 327 pages. 1977.

Vol. 579: Combinatoire et Représentation du Groupe Symétrique, Strasbourg 1976. Proceedings 1976. Edité par D. Foata. IV, 339 pages. 1977.

Vol. 580: C. Castaing and M. Valadier, Convex Analysis and Measurable Multifunctions. VIII, 278 pages. 1977.

Vol. 581: Séminaire de Probabilités XI, Université de Strasbourg. Proceedings 1975/1976. Edité par C. Dellacherie, P. A. Meyer et M. Weil. VI, 574 pages. 1977.

Vol. 582: J. M. G. Fell, Induced Representations and Banach *-Algebraic Bundles. IV, 349 pages. 1977.

Vol. 583: W. Hirsch, C. C. Pugh and M. Shub, Invariant Manifolds. IV, 149 pages. 1977.

Vol. 584: C. Brezinski, Accélération de la Convergence en Analyse Numérique. IV, 313 pages. 1977.

Vol. 585: T. A. Springer, Invariant Theory. VI, 112 pages. 1977.

Vol. 586: Séminaire d'Algèbre Paul Dubreil, Paris 1975–1976 (29ème Année). Edited by M. P. Malliavin. VI, 188 pages. 1977.

Vol. 587: Non-Commutative Harmonic Analysis. Proceedings 1976. Edited by J. Carmona and M. Vergne. IV, 240 pages. 1977.

Vol. 588: P. Molino, Théorie des G-Structures: Le Problème d'Equivalence. VI, 163 pages. 1977.

Vol. 589: Cohomologie l-adique et Fonctions L. Séminaire de Géométrie Algébrique du Bois-Marie 1965–66, SGA 5. Edité par L. Illusie. XII, 484 pages. 1977.

Vol. 590: H. Matsumoto, Analyse Harmonique dans les Systèmes de Tits Bornologiques de Type Affine. IV, 219 pages. 1977.

Vol. 591: G. A. Anderson, Surgery with Coefficients. VIII, 157 pages. 1977.

Vol. 592: D. Voigt, Induzierte Darstellungen in der Theorie der endlichen, algebraischen Gruppen. V, 413 Seiten. 1977.

Vol. 593: K. Barbey and H. König, Abstract Analytic Function Theory and Hardy Algebras. VIII, 260 pages. 1977.

Vol. 594: Singular Perturbations and Boundary Layer Theory, Lyon 1976. Edited by C. M. Brauner, B. Gay, and J. Mathieu. VIII, 539 pages. 1977.

Vol. 595: W. Hazod, Stetige Faltungshalbgruppen von Wahrscheinlichkeitsmaßen und erzeugende Distributionen. XIII, 157 Seiten. 1977.

Vol. 596: K. Deimling, Ordinary Differential Equations in Banach Spaces. VI, 137 pages. 1977.

Vol. 597: Geometry and Topology, Rio de Janeiro, July 1976. Proceedings. Edited by J. Palis and M. do Carmo. VI, 866 pages. 1977.

Vol. 598: J. Hoffmann-Jørgensen, T. M. Liggett et J. Neveu, Ecole d'Eté de Probabilités de Saint-Flour VI – 1976. Edité par P.-L. Hennequin. XII, 447 pages. 1977.

Vol. 599: Complex Analysis, Kentucky 1976. Proceedings. Edited by J. D. Buckholtz and T. J. Suffridge. X, 159 pages. 1977.

Vol. 600: W. Stoll, Value Distribution on Parabolic Spaces. VIII, 216 pages. 1977.

Vol. 601: Modular Functions of one Variable V, Bonn 1976. Proceedings. Edited by J.-P. Serre and D. B. Zagier. VI, 294 pages. 1977.

Vol. 602: J. P. Brezin, Harmonic Analysis on Compact Solvmanifolds. VIII, 179 pages. 1977.

Vol. 603: B. Moishezon, Complex Surfaces and Connected Sums of Complex Projective Planes. IV, 234 pages. 1977.

Vol. 604: Banach Spaces of Analytic Functions, Kent, Ohio 1976. Proceedings. Edited by J. Baker, C. Cleaver and Joseph Diestel. VI, 141 pages. 1977.

Vol. 605: Sario et al., Classification Theory of Riemannian Manifolds. XX, 498 pages. 1977.

Vol. 606: Mathematical Aspects of Finite Element Methods. Proceedings 1975. Edited by I. Galligani and E. Magenes. VI, 362 pages. 1977.

Vol. 607: M. Métivier, Reelle und Vektorwertige Quasimartingale und die Theorie der Stochastischen Integration. X, 310 Seiten. 1977.

Vol. 608: Bigard et al., Groupes et Anneaux Réticulés. XIV, 334 pages. 1977.